Integrated Transboundary Water Management in Theory and Practice

Integrated Transboundary Water Management in Theory and Practice

Experiences from the New EU Eastern Border

Edited by **Geoffrey D. Gooch and Per Stålnacke**

IWA Publishing
London • Seattle

Published by IWA Publishing, Alliance House, 12 Caxton Street, London SW1H 0QS, UK

Telephone: +44 (0) 20 7654 5500; Fax: +44 (0) 20 7654 5555; Email: publications@iwap.co.uk;
Web: www.iwapublishing.com

First published 2006

Index prepared by Indexing Specialists (UK) Ltd, Hove, UK
Typeset by Gray Publishing, Tunbridge Wells, UK

British Library Cataloguing in Publication Data
A CIP catalogue record for this book is available from the British Library

Library of Congress Cataloging-in-Publication Data
A catalog record for this book is available from the Library of Congress

ISBN: 1843390841
ISBN13: 9781843390848

Contents

Preface *ix*

About the editors *xi*

Contributors *xiii*

1. Introduction: identifying and solving problems in an integrated approach 1
   G.D. Gooch and P. Stålnacke
   - 1.1 Introduction to recent changes in European water policy 1
   - 1.2 Transboundary water management: a particularly problematic issue 4
   - 1.3 Water management in the new EU border regions 6
   - 1.4 The problems facing transboundary cooperation in northeast Europe 8
   - 1.5 Geo-politics in the Baltic States 9
   - 1.6 Transboundary cooperation in the Baltic States 11
   - 1.7 Environmental problems in the Baltic States 13
   - 1.8 The need (and challenges) of using an integrated approach in water management 15
   - 1.9 Integration! Integration? 17
   - 1.10 Scientific and disciplinary differences 18
   - 1.11 The MANTRA-East approach 20
   - 1.12 Scenarios as a methodology for an integrated approach to water management 21
   - References 22

2. The physical environment: Lake Peipsi and its drainage basin 25
P. Stålnacke, S. Vandsemb, T. Nõges, P. Nõges, D. Mourad, M. van der Perk, S. Nilsson, F. Hannerz, S. Langaas, E. Loigu, A. Heinsalu and B. Skakalsky
2.1 Introduction 25
2.2 Geographical features of the Lake Peipsi and its drainage basin 26
2.3 Point and non-point source pollution in the Lake Peipsi basin 29
2.4 River nutrient loads in the drainage basin 34
2.5 Water quality of the lake 40
2.6 Concluding discussion 51
References 52

3. The administrative and political environment 55
J. Johansson and L. Kalashnyck
3.1 Introduction 55
3.2 The Baltic States 56
3.3 Environmental management in Russia during the early 2000s 57
3.4 Environmental management in Estonia in the early 2000s 65
3.5 Background of the CAP 72
3.6 Problems and co-operation 80
References 81

4. Qualitative scenarios 83
G.D. Gooch
4.1 Introduction 83
4.2 Scenarios for the Russian–Estonian border 85
4.3 Discussion of the concepts and frameworks behind the scenarios 93
References 100

5. Quantitative scenarios and modelling 101
D. Mourad, M. van der Perk, T. Nõges, P. Stålnacke, M. Pihlak, E. Loigu, K. Piirimäe and B. Skakalsky
5.1 Use of models in the scenarios for Lake Peipsi 101
5.2 Riverine and source emission modelling 102
5.3 Implementation of the scenario storylines 110
5.4 Modelled nutrient emissions and loads in the 2015–2019 period under various future scenarios 111
5.5 Ecological modelling of Lake Peipsi 116
5.6 Discussion and concluding remarks 122
References 126

6. Communicating with stakeholders and the public 127
G. Roll and J.G. Timmerman
6.1 Introduction 127
6.2 Managing transboundary waters on the European Union external borders 131
6.3 Environmental information for end users 133
6.4 Development and partial implementation of an information and communication strategy for the Lake Peipsi region 139
6.5 Summary 146
References 147

7. The way ahead for transboundary integrated water management? 149
G.D. Gooch, P. Stålnacke and G. Roll
7.1 Introduction 149
7.2 Public, political, and administrative information for transboundary water management 150
7.3 Natural scientific information for transboundary water management 152
7.4 The water framework directive as a tool for transboundary water management? 153
7.5 Use of models in the scenarios for Lake Peipsi 156
7.6 Institutional frameworks and water policy 158
7.7 Conclusions and recommendations 160
7.8 Practical recommendations 162
References 166

Index 169

Preface

This book is a result of the MANTRA-East research project, the full name of which was 'Integrated Strategies for the Management of Transboundary Waters on the Eastern European Fringe'. The project was financed by the European Commission through the EU Fifth Framework Programme, Contract No. EVK1-CT-2000-00076.

About 80 scientists with widely varying backgrounds worked in the project for a period of 3 years (2001–2004). They came from thirteen different research groups and from six different countries. During the project these scientists tested ideas on each other, gathered information, analysed each other's theories and results, wrote reports, presented results at conferences and workshops, and discussed practical recommendations for transboundary integrated water management. Behind the chapters presented here lie many hours of discussions and creative contributions from this large, competent and heterogeneous research community.

We would like to thank everyone who has contributed to this book, both the 17 who have authored the chapters and all those who have contributed in other ways. The experience of working in a multi-disciplinary project such as MANTRA-East is extremely stimulating, and we are convinced that each of us has learnt much about the complexities of integrated water management through the interactions with colleagues from other disciplines than our own.

Further information and other outcomes from the MANTRA-East project can be found at htpp://mantraeast.webmedia.ee.

Geoffrey D. Gooch and Per Stålnacke

About the editors

Geoffrey D. Gooch is Professor of Political Science and Jean Monnet Chair of European Political Integration at Linköping University, Sweden. His main research interests are within the fields of International Public Policy and Political Communication, with a focus on the interaction of institutions at various levels. He is a specialist in environmental policy, and studies how changes in values, norms, and beliefs, influence the formulation of policies. He has followed developments in the Baltic States and Russia since the beginning of the 1990s.
Address: Department of Political Science, Linköping School of Management, Linköping University, SE-58183 Linköping, Sweden.
E-mail: geogo@eki.liu.se

Per Stålnacke, Ph.D. is a Senior Research Scientist at NIVA – Norwegian Institute for Water Research. During the MANTRA-East project he was employed at Jordforsk – Norwegian Centre for Soil and Environmental Research. He has long-term experience in issues devoted to integrated river basin management with particular emphasis on studies of pollutant fluxes in river basins (from source to final recipient), statistical analysis of historical environmental monitoring data, and has developed statistical models for source apportionment of pollutants and time-trend analysis. He has a strong personal interest to bridge the gaps between research and practical water management.
Address: NIVA, P.O. Box 173, N-0411 Oslo, Norway.
E-mail: per.stalnacke@niva.no

Contributors

Fredrik Hannerz
Royal Institute of Technology (KTH),
Department of Land and Water
Resources Engineering
SE-100 44 Stockholm, Sweden

Atko Heinsalu
Tallinn University of Technology,
Institute of Geology
Estonia pst. 7, 10143 Tallinn, Estonia

Jonna Johansson
Linköping School of Management,
Linköpings University
SE-58183 Linköping, Sweden

Leonid Kalashnyk
Obolonskiy Prospekt 16-E, apt. 285
04205 Kiev, Ukraine

Sindre Langaas
Royal Institute of Technology (KTH)
Department of Land and Water
Resources Engineering
SE-100 44 Stockholm, Sweden

Enn Loigu
Department of Environmental
Engineering, Tallinn University of
Technology
Järvevana tee 5, 19086 Tallinn, Estonia

Daniel S. J. Mourad
Department of Physical Geography,
Utrecht University,
P.O.Box 80115, 3508 TC Utrecht,
the Netherlands

Susanne Nilsson
Royal Institute of Technology (KTH)
Department of Land and Water
Resources Engineering
SE-100 44 Stockholm, Sweden

Peeter Nõges
Institute of Zoology and Botany,
Estonian University of Life Sciences
Võrtsjärv Limnological Station,
61101 Rannu, Tartumaa,
Estonia

Tiina Nõges
Institute of Zoology and Botany,
Estonian University of Life Sciences
Võrtsjärv Limnological Station,
61101 Rannu, Tartumaa,
Estonia

Marcel van der Perk
Department of Physical Geography,
Utrecht University
P.O.Box 80115, 3508 TC Utrecht,
the Netherlands

Margus Pihlak
Institute of Mathematical Statistics, University of Life Sciences
J. Liivi 2-513, 50409 Tartu, Estonia

Kristian Piirimäe
Department of Environmental Engineering, Tallinn University of Technology
Järvevana tee 5, 19086 Tallinn, Estonia

Gulnara Roll,
Peipsi Centre for Transboundary Cooperation
Lai 30, Tartu 51005, Estonia

Boris Skakalsky
Russian State Hydrometeorological University,
Malookhtinskiy pr., 98, 195196, St. Petersburg, Russia

Jos G. Timmerman
Department of National and International Affairs at the Institute for inland Water Management and Waste Water Treatment (RIZA)
P.O. Box 17, 8200 AA Lelystad, The Netherlands

Stine Vandsemb
Jordforsk (Norwegian Centre for Soil and Environmental Research)
Frederik A. Dahls vei 20, N-1432 Ås, Norway

1

Introduction: identifying and solving problems in an integrated approach

G.D. Gooch and P. Stålnacke

1.1 INTRODUCTION TO RECENT CHANGES IN EUROPEAN WATER POLICY

Water is of profound importance for biodiversity, and the protection of water resources a prerequisite for environmental sustainability. Water is also an element that takes little notice of political and administrative borders. However, during the early years of the European Economic Community the environment was not an area of common European Union (EU) policy. Although the European Commission (EC) began adopting measures aimed at protecting the environment as far back as 1973, it was not until after the Single European Act of 1987 that these were formally included in EU cooperation. In the period following the Danish vote on the ratification of the Maastricht Treaty in 1992, water policy was however identified by the EU member states as a major priority, and bathing water, drinking water and 'ecological waters' were seen as areas most in need of reform (Jordan 2000).

 Integrated Transboundary Water Management in Theory and Practice: Experiences from the New EU Eastern Borders edited by Geoffrey D. Gooch and Per Stålnacke. ISBN: 1843390841. Published by IWA Publishing, London, UK.

In 1995, the European Parliament's Environment Committee hosted a water 'summit' to capture the debate, at which Ken Collins, the Committee's Chairman, proclaimed that 'Our present system of a variety of European directives and decisions on water is inadequate.... Putting them all together reveals a myriad of inconsistencies, which have led to confusion and worse, non-compliance. We are paying the price for this in our water quality' (ENDS 1995). One of the most important results of the efforts to rectify such discrepancies is the EC Water Framework Directive (WFD) of 2000 (EC 2000).

The WFD represents a new approach to water management in most European countries, as it stipulates a management model based on hydrological borders and watersheds as opposed to traditional political and administrative borders. As a Directive the WFD is a form of Community Legislation that must be implemented by the national laws or regulations of the Member States within a designated time, and the WFD thus constitutes an attempt to coordinate sustainable water management in all EU member states. However, while the multi-level governance aspect of the WFD is central, 'Subsidiarity has left an indelible mark on the proposal, which leaves states to identify the relevant waters in the light of local conditions and circumstances' (Jordan 2000). The implementation of the WFD is therefore dependent on countries' national authorities, and also on the ways in which different geographical and administrative levels (European, national, regional and local) and actors work together towards a common aim. Also, despite the transboundary elements of much European water, little guidance is provided by the WFD for management of these waters.

It is now also generally accepted that water-related problems are complex and problematic, due to the fact that cause and effect relationships and abatement strategies are not independent of each other. There has however also been a long history of non-compliance with EC and EU water directives (Jordan 1999), and there are a number of difficulties that prevent the successful implementation of sustainable water management principles. For example, water pollution problems are often interrelated with environmental change and socio-economic development, and water quality issues related to surface waters are connected to both hydrological and terrestrial biogeochemical processes, including land use change and other basin-wide anthropogenic issues. Problems affecting implementation include economic, political, cultural and social issues, geographical discrepancies, shortage of financial resources, training and organisational problems, and administrative capacity.

Political and administrative values and beliefs also play an important role, as these influence both perceptions of the nature of the water-related problems to be tackled, as well as belief in the optimal ways to manage them. The National Research Council report, for example, states that 'Assessment of risks and response options tends to follow, rather than lead, political target setting, and the range of options tends to contract over time.' The problems associated with policy change are also important, and it is claimed that 'Policy is often strongly path dependent in that early decisions may constrain or determine later ones, thus making discussion of alternative policies extremely difficult at later stages.' Also that:

> Coercive sanctions have limited effectiveness in enforcing compliance, compared with carefully designed, linked systems involving rule design, information provision,

> granting of capacity and legal authority to selected actors, and transparent processes of implementation distributed among multiple formal and informal institutions. (Committee on Global Change Research 1999)

In 1999 the National Research Council highlighted a number of research imperatives for the coming decade (Committee on Global Change Research 1999). Among these imperatives were 'understanding institutions…' and 'improving methods for decision-making…'. Although this report was primarily concerned with Global Change, the highlighting of these research imperatives is relevant for many other areas of environmental politics and policy, and therefore also for sustainable water management in Europe. The report noted that while substantial work had been conducted on both international and national environmental policy instruments, and especially on the formation of international regimes, knowledge needs to be gained to be able to 'identify specific combinations of policy instruments…' and to 'identify specific international and national institutions that can effectively link the international, national and local levels…' (Committee on Global Change Research 1999). These statements stress the importance of both formal and informal institutions; that is, both organisations and norms and values, yet the role of these in European environmental post-decisional politics and implementation have not as yet been sufficiently examined (Jordan 1997).

These comments indicate that water management must integrate natural science competencies with socio-economic and policy science insights, and water policy analysts have come to increasingly recognise that managing water resources can no longer be regarded as an independent field of expertise and separate domain of public policy. It must be accepted that the interconnections between the water system, the rest of the environmental system and the human system should be studied in particular (Young and Osherenko 1993). This involves also gaining a greater understanding of human water-related beliefs and values, and the ways that these beliefs and values are influenced by, and influence, the other components of the system, as well as their effect on policy and implementation strategies.

In many cases official attention has been focused on transposing European into national laws (Ward *et al.* 1997), and a common point of departure in water management can often be found in what might be termed a legal basis, that is, it begins with legal principals of agreements and conventions. While this approach has undoubtedly contributed to a divergence of European water policy it has its weaknesses, namely that it does not take sufficiently into account the diversities in the European administrative systems that are expected to implement this legislation. As these administrations differ considerably between urban and rural, and between northern European and southern European areas, as well as being dependent on country-specific factors, what is needed is a European-wide theoretical framework for analysing the potential for the successful implementation of water management and the WFD. For it is during the implementation processes that major difficulties are most likely to arise.

Unfortunately, implementation in this field has not been extensively studied by European scholars (Jordan 1997) and there is still much work to be done in this field. Among the issues that need to be analysed are whether or not there exists a 'north–south' divide when it comes to the implementation of environmental legislation. This has been discussed by a number of scholars of European environmental policy. Ward *et al.*, for example, argue that while there are significant differences between the southern European countries (Spain, Portugal, Italy, Greece), there may yet be similarities in political culture that contribute to a common tendency to 'non-implementation' of certain aspects of European environmental law (Ward *et al.* 1997). Referring to La Spina and Sciortino (1993) Ward *et al.* claim that this may be dependent on similarities in administrative cultures which

> sanctions non-co-operative and non-compliant behaviour; administrative structures and traditions which make enforcement of regulative as well as distributive policies difficult and often random; and viscous, fragmented, reactive, and party-dominated legislative processes. (Ward *et al.* 1997)

Ward *et al.* also note that environmental management is often coupled to regional planning, and that together the demands placed upon local administrations are often difficult or impossible to meet. As an explanation for these mismatches between northern and southern European environmental implementation Ward *et al.* present two conflicting theses: a modernisation thesis based on Ronald Inglehart's theories of post-materialism, and a diverse development trajectories thesis that opposes the single linear development postulated by Inglehart. Having raised these important issues, Ward *et al.* unfortunately do not follow them up in detail.

1.2 TRANSBOUNDARY WATER MANAGEMENT – A PARTICULARLY PROBLEMATIC ISSUE

Transboundary water management presents an even greater number of challenges to politicians, planners, administrators and scientists, due to the involvement of different political and administrative systems. It has even been argued that difficulties in river basin management are only exacerbated when the resource crosses international boundaries (Wolf *et al.* 1999) (Figure 1.1). In fact, transboundary basins cover some 45% of the earth's continental land surface (excluding Antarctic) (Wolf *et al.* 1999), thus clearly indicating the global dimension of the problem.

The secretariat of the United Nations Commission for Europe states that the pollution of transboundary waters to be a widespread phenomenon in Europe, and the quality and ecosystems of transboundary waters are threatened by untreated or insufficiently treated municipal sewage, chemical pollution from industries and agriculture, seepage from old and new landfills, accidental pollution and atmospheric depositions. Damages are also caused by discharges of phosphorus and nitrogenous compounds, which stimulate eutrophication and sediment contamination, and by toxic and persistent substances (UN/ECE 1994).

Research into transboundary waters has increased, yet a substantial amount of research conducted in this field during the last decades has concentrated on water conflicts caused by lack of water, predominantly in the Middle East, Africa, India

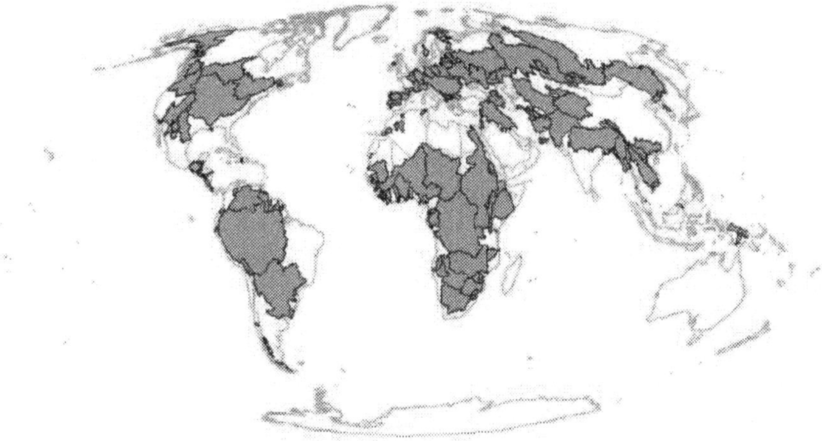

Figure 1.1 International river basins of the world. After Wolf *et al.* (1999).

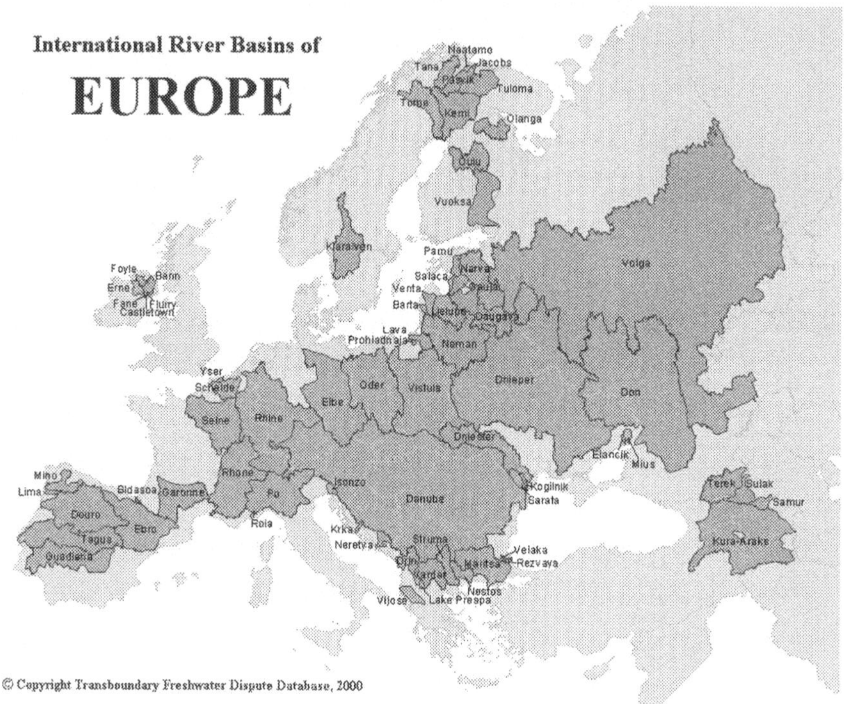

Figure 1.2 Transboundary river basins in Europe. After Wolf *et al.* (1999).

and America. Problems just as pressing exist however even in Europe today, where the difficulties encompass both water quality and water quantity. This can also be exemplified by simply mapping the number and areal coverage of transboundary basins in Europe (Figure 1.2).

1.3 WATER MANAGEMENT IN THE NEW EU BORDER REGIONS

Further complications in water management will be increased in the near future. The EU was enlarged in May 2004 by the entry of 10 new member states. Of these, eight were from the former Soviet sphere of influence, including the four countries in the northeast of Europe – Estonia, Latvia, Lithuania and Poland. The accession of the three Baltic States and Poland to the EU has moved the EU's external border eastwards, past the former 'iron curtain' (Figure 1.3).

All of the four new EU states are, even after over a decade of independence, still attempting to come to terms with the major changes that have occurred in their political and economic systems. They are also faced by problems with their neighbours, as the formation of the new border places new demands both on the member states of the enlarged EU and on Russia. It has created difficulties arising

Figure 1.3 Map of Central and Eastern European countries. *Source*: www.rec.org

from institutional interaction at regional and local levels, as well as problems associated with the implementation of national, regional and local environmental policies. Looking at the two sides of this new EU border, we can see that in Western Europe a contemporary trend has involved a shift of political power in what at first sight seems to be two opposite directions: from the national state to the EU and from the national central state level to the regions.

On closer examination, however, it can be seen that these two developments are often interconnected. They can be seen as forming different parts of a circle of redistribution of political power in the EU. It is now common to describe this system as one of multi-level governance, a structure in which varying levels of decision-making and implementation are intertwined together with the business sector and civil society. Within the Russian Federation, however, a process of re-centralisation of political power has taken place, especially after 2000. This provides a radically different framework for political decision-making and implementation of policies than the multi-level system of the EU. As regards environmental policy implementation, the current system in Russia is especially complicated and presents a significant challenge to EU–Russian cooperation.

In the new EU states and Russia, employment in the agricultural sector is also considerably higher than the average for the former EU 15 countries. The results presented in this book from the MANTRA-East project show that agriculture is a major source of diffuse pollution, even though the decrease in agricultural production that occurred at the beginning of the 1990s has resulted in substantially lower levels of artificial fertilizer use. The introduction of EU Common Agricultural Policy (CAP) funds, even at the introductory level now planned for these countries, will however in turn once again lead to significant changes in agricultural practices and production. The new EU states are still attempting to come to terms with the changes that have occurred in their systems, and the administrative structures concerned with environmental management are still in a state of reorganisation and development. The formation of the EU's eastern border towards Russia has placed increased demands on the new member states, as well as on the EU's administrative capacity. It also entails changes in EU–Russian relations. These difficulties are significant at the interstate level, and they are especially pronounced in the case of institutional interaction at regional and local levels, where problems arise in the implementation of national, regional and local environmental policies. The nature of these problems is both political and administrative. It is therefore important to examine those aspects of cooperation that affect water management policies and that may lead to difficulties in the political and administrative management of transboundary water management.

The WFD stipulates a management model based on hydrological borders and watersheds as opposed to traditional political and administrative borders and the accession of the Baltic States to the EU has placed new demands on institutional interaction at regional and local levels, and on the implementation of water management policies. The chapters in this book have therefore examined those aspects of cooperation that affect environmental policies and lead to complications in the political and administrative management of transboundary environmental relations. They have analysed the potential problems facing this region through the use of qualitative and quantitative scenarios, which present possible future developments in a region that is undergoing major economic and social changes, and where conditions can be best characterised as uncertain. These scenarios (Chapters 4 and 5) attempt to provide indications of how future developments on the Baltic–Russian borders may occur, and they are an important component in a process that combines qualitative and quantitative aspects of environmental policy.

1.4 THE PROBLEMS FACING TRANSBOUNDARY COOPERATION IN NORTHEAST EUROPE

It can be argued that the most important process that has taken place in Europe after World War II is voluntary integration in the form of the EU. It should be remembered, however, that integration is not a steady and level process, that it can be voluntary and involuntary, and that it is not irreversible. Europe also consists of more than just Western Europe, and while the EU has in fact developed and expanded since the 1950s, albeit in an uneven and sometimes erratic fashion, the countries of Central and East Europe first experienced the forced integration of the USSR and COMECON, and, from the 1980s on, the process of disintegration of these constructs. We therefore have not one but at least two forces that have formed Europe as we know it today, and the effects of these two processes will determine transboundary relations for many years to come. These forces have created three different forms of transboundary relations in the new Europe. Firstly, there are transboundary relations between the existing members of the EU, and between the EU and non-EU Western European countries such as Switzerland. These are regulated by the EU Treaties, and by established practices. Secondly, there are transboundary relations between the EU and the applicant states of the Balkans, and between the applicant states themselves. Thirdly, there are transboundary relations between ex-Soviet or COMECON countries and the Russian Federation that, as the enlargement of the EU has moved EU borders eastwards, has created new EU cross-border relations with Russia, Belarus, Ukraine, etc.

Starting at the regional and local levels, it can be seen that one of the most important trends in regional and local administration during the last decades has been decentralisation (SOU 1998). There seem to be a number of driving forces behind the trend towards the devolution of central political power to sub-central governmental levels. These include a belief in the greater efficiency of regional and local political and administrative units, ambitions to increase and revitalise democracy, and the needs of central governments with dwindling budgets and increasing costs to pass on the responsibility for implementation policies to other lower levels. Decentralisation usually creates positive connotations in Europe and is seen to be synonymous with increased political participation, better possibilities to influence decision-making, greater local autonomy, more effective organisations and benefits for democracy. It is therefore seen as a necessary realignment of power from large central institutions to institutions at a lower administrative level. The move towards decentralisation can be seen in many European countries, and especially in countries such as France, Italy, Portugal, Great Britain and the Nordic countries (SOU 1998, p. 69).

In Western Europe this recent interest in decentralisation has often been connected with regionalism. A number of factors have influenced this renewed interest in regions. Firstly, economic development during the last decades has in many countries been erratic, and negative trends have led to a renewed interest in the conditions that affect long-term economic growth. Secondly, the EU has contributed to an increase in the importance of regions for the member states' economies. Significant parts of the EU's programmes, and a large percentage of its budget, are aimed at improving regional activities and development. The free movement goods, services, investments and labour have led to a decline in the importance of national boundaries as far as trade is concerned (Hallin and Malmberg 1996) and regional economies are expected to play an increased role in the future (Krugman 1996, p. 17). The Four Freedoms have not however led to a similar decline of the role of borders as far as culture and identity are concerned. It should be remembered that while the EU has achieved much in the way of economic integration, and perhaps less as far as political integration is concerned, it still has not achieved any great measure of cultural integration. Identity in Europe is still very much a question of national identity, and instead of a trend towards a European identity, the move seems to be more towards regional and local identities.

At the same time as these changes have been occurring in the EU, much more radical transitions have been taking place in the Central European Countries (CEC) and Russia. The problems caused by the collapse of the CEC's economies in the beginning of the 1990s were to a large extent the result of the sudden loss of the Soviet market, and in some countries problems with ethnic minorities. After the denial of nationalism during the Soviet period the rise of ethnic conflicts and nationalism in the 1990s has created problems, especially within the Balkan and Baltic regions, and between ex-Soviet countries with ethnic Russian populations and the Russian Federation. The CEC and Russia are also experiencing problems caused by the lack of experience and competence of many of the authorities that are expected to formulate and implement policies. After the fall of the iron curtain, hierarchical and authoritarian forms of government have officially gone out of fashion in many parts of the CEC, and attempts have been made to replace these with efficient and democratic forms of politics and administration. This is not an easy task and the problems facing these countries are extremely complex. Cross-border and interregional cooperation between 'old' EU countries and the new member states, and between the new member states and Russia are therefore not only faced with economic and political instability, but they also need to come to terms with the problems of countries with radically different political and administrative traditions and cultures, each trying to solve different sets of problems, often by different means.

1.5 GEO-POLITICS IN THE BALTIC STATES

The Baltic States lie on the east coast of the Baltic Sea, in the western part of the East-European Plain (Figure 1.3). The countries are mostly flat and large parts of them consist of lowlands and predominantly rural communities. Among the countries of former Eastern Europe the Baltic States hold a special position as they were integrated directly into the Soviet Union as Soviet Republics in the 1940s. Other countries of Central and Eastern Europe were dominated by Communist systems but not directly a part of the USSR. After the break up of the Soviet Union many of the former Soviet republics became members of the new Confederation of Independent States (the CIS). Estonia, Latvia and Lithuania did not. Despite a number of similarities between the three Baltic States they cannot however be bundled together as a block. Differences in history, population, economies, linguistic and culture are significant, and it is often easier for these countries to cooperate with other non-Baltic countries than with each other. Russian was the common language used in communication between the three countries during the Soviet period but resentment against the Soviet occupation has left many people in the Baltic States reluctant to use Russian, and therefore without a common language.

The issue of language and ethnic groups is in fact one of the problems that Estonia and Latvia faces today. As Glenny (1993) has pointed out, during the Communist regime ethnic conflicts were not admitted, as nationalism was seen to be something that only belonged to capitalist society. Ethnic conflicts did exist however, such as those between ethnic Germans and Vietnamese workers in East Germany, and the anti-Semitism that was found in several East-European countries (Glenny 1993). During the 1990s nationalistic sentiments became apparent in many of the former Eastern European countries, sentiments that were in part held in check by these countries wish to become members of the EU and the related need to conform to EU norms and policies. Citizenship legislation that came into force in Estonia in 1992 and in Latvia in 1994 limited the rights of the ethnic Russian 'minorities' and made naturalisation a complicated process.

Following the break up of the Soviet Empire at the end of the 1980s and independence for the Russian Federation in 1992, protection of the Russian community outside of Russia became an important question on the political agenda for the new Russian Federation. Many of the countries of the former Soviet Union found themselves with significant ethnic Russian populations and about 60 million former Soviet citizens suddenly found themselves 'abroad'. Of these, about 25 million were Russians and approximately 35 million belonged to other nationalities (Sakwa 2002). This means that about 17% of the total Russian population of the former USSR came to live in the 'near abroad' (Sakwa 2002); of these, almost 500,000 lived in Estonia and 800,000 in Latvia. While it may not be technically correct to equate the USSR with Russians, the sentiments of many former Soviet citizens were that the Soviet nation never existed, and that the Soviet state was seen as an extension of the Russian nation (Bremmer and Taras 1997). The struggle for independence in Latvia and Estonia had assumed an identity centred on culture, language and nationalism, and this led to the emergence in the beginning of the 1990s of social and political conflicts based on ethnic identity. The ethnic conflict of the early 1990s resulted in, and was aggravated by, the disputes over citizenship rights in the two countries, in which ethnic Russians were seen by themselves and by Russia, as discriminated. This was most obvious in the conflict between the new national government in Tallinn and the predominantly ethnic

Russian city of Narva on the Estonian–Russian border. In Latvia the situation was aggravated by the prolonged conflict over citizenship rights, and the lack of political influence of the ethnic Russia population (Melvin 1995). It should be noted, however, that relations between ethnic Latvians and ethnic Russians were not always conflictual. Far from it, the move to independence in these countries could not have succeeded without the neutrality or support of non-Latvians, as ethnic Latvians constituted only 52% of the republic's population in 1989 (Muiznieks 1997). Also, intergroup relations are not uncommon; in 1988, for example, one-third of marriages in Latvia were between Latvians and non-Latvians (Melvin 1995). In Latvia, ethnic minorities at the end of the 20th century accounted for nearly 40% of the population, and of these, over 600,000 people did not have Latvian citizenship; a significant proportion of these were former citizens of the USSR who had no citizenship at all (European Communities 1997).

Relations between Estonia, Latvia and Russia were (and still are to a certain extent) also complicated by the fact that the vast majority of the new political elite are either native Estonian or Latvian speakers, with Estonian and Latvian being the official languages of these countries. There are however still parts of these countries, and also many areas in the main cities, where the Estonian and Latvian languages are hardly spoken at all (Dawisha and Parrott 1997). The demands made upon language proficiency in Estonian and Latvia as a prerequisite for work in the bureaucracy led to a rapid decrease in the number of non-Estonians and non-Latvians working as civil servants, and to an influx of new, mostly young and sometimes inexperienced, administrators. Ethnic Russians could therefore sometimes find themselves in situations in which they were dependent on decisions taken by civil servants who were unaware of their problems at best, or directly opposed to their presence in these countries at worst (Melvin 1995). Despite recent improvements this can still constitute a major problem for parts of the countries' populations.

1.6 TRANSBOUNDARY COOPERATION IN THE BALTIC STATES

Moving on to Baltic–Russian cross-border cooperation, it should be noted that the ways in which administrations function is the result of a large number of factors; these include traditions, resources and levels of competence. There are also near relationships between the norms of political systems and the ways in which these political systems organise administrative structures (Blondel 1995). Norms are a part of the political culture of a country, and countries each have specific ways of making political decisions and implementing those decisions (Petersson 1995). In the case of the Baltic States and Russia these differences may prove to be the hardest problem to solve, for as Nordlöf-Lagerkrantz and Lagerkrantz (1993) pointed out in the early 1990s, the heritage of the Soviet era was (and still is) strong. When two different political and administrative cultures attempt to work together differences in bureaucratic norms and standard operating procedures between organisations can create serious problems. These differences may be even more significant than the political landscape of the municipalities. Administrative cultures can play a greater role than political loyalties. Closely associated with the issue of norms and political culture is the question of mentality, and how to change the ways in which people think.

In an ideal situation the administration would be able to guarantee the legal rights of citizens, to be able to objectively process cases, and also to follow political decisions. Unfortunately many of the countries of the former Eastern Europe lacked the administrative capacity necessary for the successful implementation of new policies (Toresson 1995), and some still do, as has been noted in the EU Agenda 2000 report. This lack of satisfactory administrative structures is to be found in a number of Baltic municipalities and is even more acute on the Russian side of the borders. It creates major problems for all forms of transboundary cooperation. Even in the Baltic States, which have come far since 1989, the EC has noted that they still have problems with the administrative structures necessary to successfully implement environmental and legal issues. It should be remembered that as former Soviet republics the Baltic States have been forced to build upon a legacy of centrally steered administrations, and governmental structures and legal authorities that were often seen primarily as ways of guaranteeing political control. The experiences of the last 50 years created administrations in which civil servants were not sufficiently experienced in independent decision-making and implementation. These administrations needed to develop the necessary competence for the new forms of work, and this has been a major hinder for the implementation of policies in these countries. While the competence of the civil service has been raised considerably, and ways to improve administrative structures have been initiated, there is still much to be done.

Despite the problems facing cooperation, new transboundary networks of actors consisting of groups of organisations and/or people who have a common interest in a policy's development and implementation are being created. These networks consist of politicians and administrative officials from both high and low levels of an administration together with, in some cases, pressure groups and representatives of private businesses. This cooperation is most successful when it involves a number of the main political and administrative leaders in the participating regions as this demonstrates the importance of the project. The role of political leadership is still pivotal in many parts of the former Soviet Union. In theory these networks can be quite large (Petersson 1998) but they may also only consist of a few key people.

To a certain extent the difficulties associated with the implementation of cross-border regional cooperation is a problem of both goals and methods. Civil servants and politicians in the Baltic States can adopt approaches that are radically different from the top-down authoritative methods of the Communist era that still exist in many parts of Russia. Two different traditions can also meet in the Baltic States themselves. First, a traditional way of looking at an administration as a hierarchical, technically competent organisation that is able to rationally solve problems, an approach with some similarities to Max Weber's (1864–1920) view that an effective, efficient and predictable administration was a prerequisite for functional public service (Held 1987). Ex-Communist versions of hierarchical administrative structures are common in post-Soviet countries, and even if their main aims have changed since the fall of Communism, in many authorities their organisational methods and structures have not. Pitched against this model is the network model based on consensus seeking and discussion. A major problem for cross-border cooperation is therefore to reach agreement on standard operating procedures, for as Hill has noted 'the policy-making process will be influenced by constitutions, rules, political structures and standard operating procedures' (Hill 1997, p. 123). In order for the participants to successfully work together there has to be some kind of consensus on goals and methods. If there are differences, the question is, who will change?

In effect there are often also two conflicting methods of policy implementation representing rational and incremental models. In the rational implementation model a finalised plan or vision is seen as one of the first steps in the implementation process. According to this model information on alternative solutions needs to be collected and an evaluation made of the possible alternatives. The officials responsible for implementation should then follow these directives.

The incremental model prefers to address each smaller problem as necessary. There are two arguments that can be used against cross-border cooperation based on the use of rational decision-making models in the Baltic States and Russia at the present time. One is that there is a tradition of formulating grand plans in the former Soviet Union. These plans then might or might not be adhered to. Continuing this tradition of formulating plans that are then not adhered to is not conductive with efficient implementation. The second difficulty with this model is that rational implementation demands considerable resources, a reasonably predictable future scenario and an absence of major conflicts between different interests. None of these can be said to exist today. The main argument against the incremental model, on the other hand, is that it is more difficult to achieve the major changes that are necessary through the piecemeal processes that the model advocates.

To sum up, the main problems facing cross-border cooperation in northeast of Europe are:

(i) a lack of funds to cover the costs of the work,
(ii) intercultural differences in administrative cultures and communication,
(iii) low levels of background knowledge that often intensifies these problems as does lack of experience of international cooperative projects,
(iv) different administrative cultures and experience of the participating authorities.

1.7 ENVIRONMENTAL PROBLEMS IN THE BALTIC STATES

The severe eutrophication problems of the Baltic Sea have focussed attention and placed major emphasis on nutrient inputs via rivers and others sources. Nutrient losses from agriculture are considered to be responsible for the major input of nutrients to surface waters and maritime areas in almost all the Nordic and Baltic countries (Stålnacke 1996; Lääne *et al.* 2002). Although both the predominant nutrient transformation processes in agricultural soils and the main transport processes for nutrient losses at a global scale are well documented, it is hard to predict the level of nutrient losses from agricultural landscapes and how different measures may influence these losses. The large spatial and temporal variability in nutrient losses, water discharge, weather conditions and farming practices also make it very difficult to distinguish between differences due to natural causes and those resulting from agricultural activities.

Agriculture in the Baltic States has undergone dramatic changes since these countries' independence in the early 1990s. The years 1990 and 1991 constitute the starting point of an almost total collapse in agricultural production due to the disintegration of the former Soviet Union. The consequences of this collapse were immediate, and included, among other things, dramatic decreases in nutrient inputs. Mineral fertiliser use decreased by 90% and the number of livestock decreased by 50% from previous levels. At the same time, large areas of arable land were taken out of production. Such abrupt and significant land use change has hardly been recorded in modern European agricultural history. Compared with the Nordic countries (Sweden, Norway, Denmark and Finland) the majority of the lowest N losses to surface waters in agriculture occurred in the Baltic countries (Figure 1.4).

Plot experiments have demonstrated that the amount of fertilisers applied to agricultural land may have a significant impact on the loss of nutrients from soil to water. In addition, there is considerable evidence that agriculture in general, or specific farming practices, can have a pronounced effect on the output of nutrients, especially nitrate, from entire river basins. Several investigators have found marked correlations between nutrient levels or concentrations in river water and the propor-

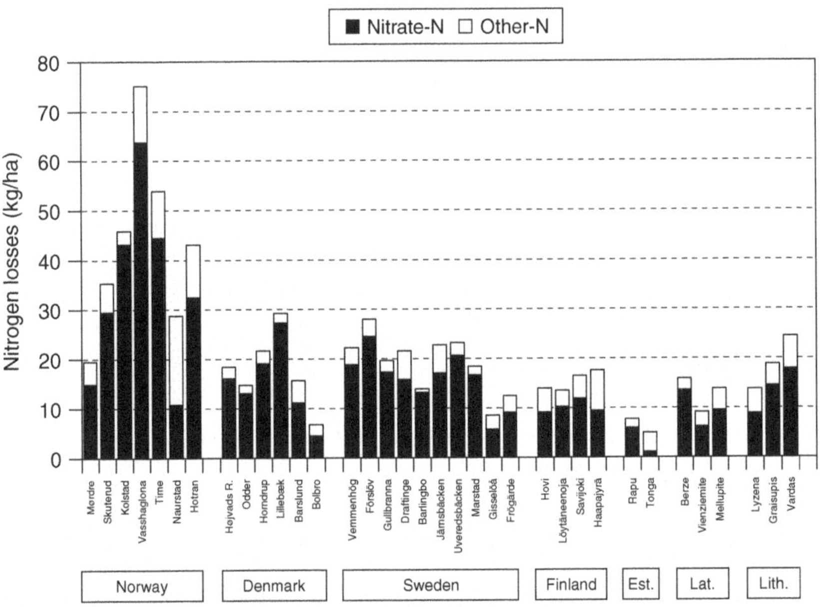

Figure 1.4 Mean annual losses of nitrogen from 35 agricultural catchments in the Nordic and Baltic countries, 1993–1997. After Vagstad *et al.* (2004).

tion of agricultural land in the studied river basins. Furthermore, examination of long time series of nutrient concentrations in European rivers has revealed upward trends in nitrate levels that have paralleled the increased use of nitrogen fertilisers, whereas a weaker relationship with agriculture has been noted for temporal trends in PO_4-P (Grimvall *et al.* 2000).

The literature suggests that the extensive decrease in agricultural intensity that began in the late 1980s in Eastern Europe has led to only a slow and limited reduction in nutrient levels in many rivers (Stålnacke *et al.* 2004). This implies that there will not necessarily be immediate responses to large cuts in nutrient inputs. This poses the important question of how long it takes for decreased inputs or emissions in a drainage area to influence the output through the mouth of the river under consideration. However, attention should also be focused on the risk of an increased riverine export of nutrients, since that export is controlled not only by emissions, but also by retention of nutrients in terrestrial and aquatic systems as well. Budget calculations for entire European river basins, and the sub-basins thereof, have demonstrated that up to 80% of the phosphorus and nitrogen released to water is either trapped in sediments or removed by denitrification during transport from the source to the mouth of a river (e.g. Behrendt 1996). Unfortunately, there is no guarantee that this retention can be maintained if Western European solutions and technologies in full are copied by Eastern Europe. This issue is further discussed in Chapters 2 and 5.

1.8 THE NEED (AND CHALLENGES) OF USING AN INTEGRATED APPROACH IN WATER MANAGEMENT

Policy-makers and water managers have traditionally regarded water quality issues as primarily originating in one specific cause: they have often taken the view that there is usually one straightforward solution to the pollution problem. Today, however, it is generally accepted that water-related problems are far more complex and problematic, and that simple cause and effect approaches are insufficient. It is now realised that abatement strategies must adopt a different approach, in which problems originating in different spheres are seen as interdependent. For example, local pollution problems may be interrelated with environmental change and socio-economic development to such an extent that a single disciplinary or sectoral approach can no longer provide a satisfactory solution. Another example is that water quality issues related to surface waters are now being correlated to both hydrological characteristics and terrestrial biogeochemical processes, including land use change and other basin-wide anthropogenic issues. Yet another aspect is the conflict between social and economic development on the one side and environmental and pollution concerns on the other. Water policy analysts have therefore increasingly come to recognise that managing waters can no longer be regarded as an independent field of expertise and a separate domain of public policy. It is now accepted that the interconnections between water systems, other aspects of environmental systems and human systems are extremely important areas of study. Thus, it has become apparent that water and river basin management should be based on an integrated approach, involving planners, scientists, policy-makers, end users and the public. Even though integrated water resource management (IWRM), integrated water management (IWM) and integrated river basin management (IRBM) are currently said to be practised in many regions worldwide, we still have not been able to 'solve' pollution and water management problems. Examples of 'success stories' are also unfortunately difficult to find. In the EU member states, new and old, the EU WFD will become a central tool for future environmental water management and legislation of river basins. However, while the EU's environmental acquis and management doctrines have been adopted by the new member states, several studies have recognised that the environmental problems being faced in these states are far more severe than in most of the EU 15 member States. Thus, there is a need for truly integrated and functioning basin-wide cooperative schemes for water management. IRBM is central in the EU WFD, and the implementation of this directive requires comprehensive interdisciplinary analyses (Schultz 2001). In this respect, the experiences gained in the research presented in this volume from the MANTRA-East project (integrated strategies for the management of transboundary waters on the Eastern European fringe – the pilot study of Lake Peipsi and its drainage basin) should prove to be valuable. The project was initiated to facilitate the search for integrative strategies and integrative examples. The pilot study was Lake Peipsi, the largest international lake in Europe, and a water body shared by one new EU member state (Estonia) and one non-EU state (Russia). The water catchment area also includes parts of Latvia. It is therefore of high relevance for the future environmental management of transboundary waters on the new EU border regions.

Figure 1.5 In the MANTRA-East project, various field investigations were carried out as a supplement to the collection of existing data and information. Here, sampling from the research vessel '*Limnoloog*' is carried out in connection to a joint Estonian and Russian expedition in August 2001. (Photo: Tiina Nõges.)

1.9 INTEGRATION! INTEGRATION?

Integration is the buzzword of today in water and river basin management, often used to demonstrate that the research being conducted is up-to-date. It is also used to distance approaches from traditional methods and conceptualisations. The term is however commonly used without further and deeper consideration. Far from being an accepted and easily understood notion around which there is consensus, the entire concept and exact definition of IWM, IWRM and IRBM are in fact widely debated. They are also often ambiguous, and a unanimously agreed definition of the concepts has yet to appear. In an attempt to remedy this, the Technical Advisory Committee of Global Water Partnership has adopted the following definition:

> IWRM is a process, which promotes the coordinated development and management of water, land and related resources, in order to maximise the resultant economic and social welfare in an equitable manner without compromising the sustainability of vital ecosystems.

Key points here are process, coordinated, and the important relationship between sustainability and economic and social welfare. The definition does not, however, give us much indication of how this coordination (and integration) is to be achieved. According to Mostert IWRM, as well as IWM and IRBM, five main characteristics may cause complications and problems (Mostert 1998): multifunctionality (e.g. fishing, farming, water supply), user interests and conflicts, multiple managers at different levels (e.g. local, regional, national), asymmetric power relations (e.g. up- and downstream users and managers), and technical complexity. The distinction between 'integrated' and 'traditional' management of water, river basins or water resources to a large extent relies on the scope and sphere of operation of orientation. Whereas 'traditional' is typically sector oriented (water supply, irrigation, hydropower, etc.) and focused on satisfying the perceived demands within each sector, 'integrated' attempts to take a cross-sectoral approach and also focus on the management of water, or in terms of water resource management on the demand, supply and use of water. In order to develop the distinction between traditional and integrated approaches, we can examine the forms of concept polarities that exist, and look at their possible forms of integration. Table 1.1 presents a selection of these; more could of course be added.

It becomes clear that integration can have multiple meanings and that the interpretation of the concept depends on the perceptions of what it is that should be integrated. While the polarities shown above all present significant problems, it is perhaps within the final distinction, that between social scientists and natural scientists, the most difficult obstacles to integration become apparent. Despite efforts by various institutes to develop interdisciplinary research into water and environmental issues (e.g., at the universities of East Anglia in the UK and Linköping in Sweden) the results can often be characterised as more multidisciplinary than interdisciplinary. The problem at hand is to develop a cross-disciplinary dialogue while at the same time retaining the distinctive scientific depth of the individual disciplines. In the following section we will examine the epistemological differences between the disciplines involved in water management, and attempt to identify the main

Table 1.1 Examples of polarities and examples of forms of integration.

Polarities	Examples of forms of integration
Economy vs. ecology	Environmental economics
Sector vs. sector	Environmental planning and cooperation between economical sectors
Nation vs. nation	Transboundary river basin management
National authority vs. municipality	Multi-level government
Government vs. non-governmental organisations	Governance
Policy-makers vs. stakeholders/ the public	'Bottom-up' policy-making
Scientists vs. policy-/decision-makers	Scientific communication strategies
Land vs. water	Land use change/nutrient loss models
Surface waters/groundwater vs. coastal waters	Coastal zone management
Water quantity vs. water quality	Water boards
Upstream vs. downstream users	River basin management
Large-scale vs. small-scale approaches	Regional perspective
Water supply vs. wastewater management	'Grey water' and use of wastewater as irrigation water
Social scientists vs. natural scientists	Multidisciplinary and/or interdisciplinary approaches

Source: Gooch (2003).

barriers to constructive interdisciplinary research. Awareness of these differences can provide a starting point for the development of truly integrated approaches.

1.10 SCIENTIFIC AND DISCIPLINARY DIFFERENCES

Each scientific discipline answers epistemological questions in a different way. According to Losee (1980) the central issues are:

(i) Which characteristics distinguish scientific inquiry in this discipline from other forms of investigation?
(ii) What procedures should scientists in this discipline follow?
(iii) What conditions must be satisfied for the explanation to be correct?
(iv) What is the cognitive status of scientific laws and principles?

The debate is ancient and in order to clarify the modern controversy on the best ways to analyse and improve water management we need to begin with the foundations of the debate, and with the ideas of the classical Greeks. Plato's idealism provides radically different answers to the question of knowledge than Aristotle's inductive/ deductive empirical methodology. While Plato's philosophical approach dominated in Europe prior to 1150, the translation of Aristotle's writings on science and scientific method during the 12th and early 13th centuries provided scholars with new insights. The scientific debate diverged and by the early 19th century the central issues were inductivism vs. the hypothetico-deductive method (Losee 1980). While the traditional mechanical view of nature was sometimes questioned, the view that dominated was 'scientific materialism', one that 'presupposes the ultimate became the ultimate fact of an irreducible brute matter, or material, spread through space in a flux of configurations … senseless, valueless, purposeless' (Whitehead 1926, p. 22). Although some philosophers of science such as Kuhn claim that science is not a steady, cumulative acquisition of knowledge, but instead a series of interludes punctuated by intellectually violent revolutions (Kuhn 1970), positivist philosophy still dominates the natural sciences. This is a philosophy based on casual laws of phenomena, derived from observation (Outhwaite 1987). In the social sciences, however, central issues include the debate on the importance of structure vs. agency, on meaning, and on the construction of social reality. Winch, for example, stresses the importance of meaningful behaviour for the study of human actions and states that 'the analysis of meaningful behaviour must allot a central role to the notion of a rule' (Winch 1958, p. 51). The concept of rules in the form of informal institutions such as norms and values is now a central issue for the social sciences. The differing traditions of the natural and social sciences obviously express themselves in a number of ways. One important aspect for the case of transboundary water management is the different attitudes to the question of open or closed systems. While natural scientists, taking an empirical, positivist point of departure, are aware of the impossibility of a closed system outside the laboratory (an even within it now that quantum physics has been accepted), they are often willing, for the purpose of prediction, to adopt an approach that postulates just such a closed system. Social scientists, on the other hand, usually believe that 'because social systems are intrinsically open and cannot be artificially closed, our criteria for the empirical testing of social theories cannot be predictive and so must be exclusively explanatory' (Bhaskar 1989, p. 5). Within water management we can therefore identify a number of significant epistemological differences between the scientific disciplines involved. The nature of the system (closed or open) being studied is one. Another is the nature of the object of study, of whether it consists of brute matter or meaningful behaviour. A third is the difference between the procedures (quantitative/qualitative, inductive/deductive, etc.) considered correct for scientists to follow when conducting scientific enquiry. Social Science is also, in general, characterised by heterogeneity, and David Braybrooke has identified three sides of social science that are important for water management (Braybrooke 1987). These are the 'naturalistic' (positivistic) view, the 'interpretive' view and the 'critical' view. The 'naturalistic' approach selects problems and utilises methods that have much in common with those predominant in the natural sciences. Proponents of this school seek casual regularities and prediction. According to the 'interpretive' view, on the other hand, the methods used by the natural sciences (and by the 'naturalist' approach) are not applicable to the study of social phenomena. Rules and meaning are instead important concepts for this approach. Finally, the 'critical' approach looks for underlying power relations, and seeks to explicate the interests of the actors and structures. Social scientists working within these schools tend to identify different problems, and to study them with different methods, and some will find it easier than others to work with natural scientists in water management issues.

When it comes to IWRM, IWM or IRBM the natural science perspective usually focuses on the different fundamental processes (biogeochemical) in river basins and water systems. Thus the natural science perspective can potentially contribute to basic information, for example, in river basin management plans. A limitation is that natural scientists focus on the object, the river basin itself, and less on the management process (Mostert 1999). In addition, the national or international character of river basins is often not important in this perspective (Marty 1997). Social scientists, on the other hand, and especially disciplines working with policy and decision-making, have their starting point in the social sphere and not the river basin (Mostert 1999). Thus the social science perspective may contribute to the understanding of the structure, ideas and strategies of actors and the 'management' of the decision-making process (Klijn and Koppenjan 1997). In contrast to natural science, social science approaches can also apply equally to national and international river basins, and especially to the differences in management at various administrative levels, as well as concerted management of transboundary river basins, and harmonisation of IRBM policies (Mostert 1998). One main limitation with the social science perspective is that IWRM and IRBM are not only about actors and scales but also about the river basin, and especially the interaction between actors and river basins (Mostert 1999). It is therefore necessary to combine various perspectives on IWRM and IRBM. As experience has shown, this is however far from straightforward. The ways in which attempts were made to overcome the problems inherent in interdisciplinary research are described in the next section.

1.11 THE MANTRA-EAST APPROACH

The MANTRA-East project had three main objectives. The first of these was to evaluate the applicability of the EU WFD in the future EU border regions. This included an assessment of the state of eutrophication (e.g. ecological status) in lakes and river basins, as well as the development of strategic lake and river basin tools for source apportionment, retention and time trends in nutrient loads. The second objective was to develop methods to improve communication and utilisation of scientific information. The third objective was to develop institutional mechanisms and policy instruments for decision-making under conditions of transition and uncertainty. The first objective was primarily the domain of the natural scientists in the project, the second objective was mostly managed by geographers and information specialists, and the third objective was the responsibility of the social scientists and policy analysts. The underlying rationale behind this model was that scientific information produced by the natural scientists was needed in the policy process, as well as knowledge of the policy process (decision-making and implementation) itself. This information needed to be communicated to the actors in the policy process, as well as to stakeholders and end users. The problem facing the project was that each objective was based on, and produced, different forms of knowledge, or, to put it in another way, each objective was dependent on a different scientific paradigm (Kuhn 1970). While respecting the different disciplinary approaches and methods, it was necessary to find a way to integrate the knowledge produced, and to create a fertile cross-pollination of ideas, concepts and hypotheses. The natural science approach was based on quantitative methods and results that could be analysed through statistical methods and mathematical models. The communication of scientific knowledge was studied through a combination of information cycle methodology and qualitative analyses of communicative processes. The policy process, the involvement of stakeholders and public participation were analysed using a heuristic approach in which qualitative interviews, surveys, focus groups, text analysis and institutional analysis were combined. The experience of the team was that understanding of each other's epistemological viewpoints could, and was, increased by continuous discussions of the nature of the problems and the best ways to solve them. In this way the project adopted a problem-based methodology, attempting to elucidate the central aspects of water management from various viewpoints. The difficulties with the approach were, after an initial period, reduced and resulted in an increased understanding of differing scientific beliefs. The next step was to develop an approach that was synergetic, and which resulted in a greater understanding of the problems of water management than if each discipline had simply presented individual results which were then combined in a final report. The approach chosen was that of qualitative–quantitative–qualitative scenarios. These are introduced in the next section and described in detail in Chapter 4.

1.12 SCENARIOS AS A METHODOLOGY FOR AN INTEGRATED APPROACH TO WATER MANAGEMENT

During recent years the use of scenarios for environmental policy-making has attracted considerable attention from both the scientific community and policy-makers (see, e.g., Greeuw *et al.* (2000), and for business aspects, Ringland (2002)). Many of these applications have however either focused on larger spatial scales, such as countries (Kahane 1997), or have utilised relatively well-known cases where information, while not comprehensive, has been comparatively readily available (Greeuw *et al.* 2000). As regards conditions in transition and post-transition countries, scenarios have not yet been so widely used. Scenarios are not precise predictions of the future (Porter 1985); they should be seen as simulation tools, as a technique similar to, but different from, models (Jouvenel 2000) as scenarios attempt to construct general representations of possible alternative futures. Alcamo (2001) identifies five main elements of scenarios: these are a description of the changes that may take place; of the main factors and driving forces that influence these changes; a definition of the beginning year of the scenario, the base year; the selection of the time frame for the scenarios and the adoption of time steps and the construction of the storylines, which are narrative descriptions of possible futures. Scenarios can be especially useful in environmental assessments as they can provide a picture of future alternative states of the environment, and this is an important aspect, as a major problem facing policy analysis is the need to manage and analyse vast amounts of information, a task that is often overwhelming (Hill 1997). Scenarios, through their simplification of this information, can make this process easier. Basically scenarios can be qualitative or quantitative, explorative or normative. In the MANTRA-East project, qualitative scenarios, 'storylines' or 'scripts', were first developed (Gooch 2003); these were narrative texts that were representations of possible futures. They did not contain detailed numerical information, and could be written in an easily understandable manner for policy-makers, stakeholders and the public. Utilising past and present trends, the qualitative scenarios attempted to see which factors might lead to likely futures. The qualitative/explorative scenarios were then used as input into computer-generated models that can best be described as quantitative/explorative in the first stage, and quantitative/normative in a second stage (see Chapter 5). According to the project methodology, the results of the quantitative models were then used to modify the qualitative scenarios, and to develop qualitative/normative scenarios that provided grounds for policy recommendations. This combination of storylines and set of model calculations is similar to that used in the 'World water vision scenarios – the world water situation in 2025' (Cosgrove and Rijsberman 2001). In the MANTRA-East qualitative scenarios, different levels of economic development and cooperation were used as determinants and three alternative future scenarios complemented the traditional 'business-as-usual' scenario. However, as the project was concerned with a border region in which two countries with differing conditions interact, it was necessary to construct two different groups of scenarios, one for each country. While these groups of scenarios were based on similar driving forces, the results of these forces differed according to the specific conditions in the countries. In order to construct regional scenarios, these country-specific scenarios were then combined. As the number of possible alternatives was considered excessive, only a limited number of regional scenarios were constructed. A problem that had to be faced was that while the focus of the study was on transboundary interaction at a regional or sub-regional level, sufficient material was not available to allow the construction of qualitative scenarios at this geographical level. Instead, possible future trends at national and sub-national levels were used to construct the regional and sub-regional scenarios. The time period chosen was 12 years, from 2003 to 2015.

Seven factors were identified as central for the scenarios: these were population, agriculture, fishing, economy, politics, administration and environment. As in the IMAGE scenarios (Alcamo and Nakicenovic 1998) the primary driving forces included population and economic growth rates. As the span of factors was relatively large, it might be claimed that even at this stage the qualitative scenarios were integrated, as they included economic, social and environmental factors. However, the qualitative scenarios were not seen as an end in themselves. Instead, they were used as an input to the quantitative scenarios, thus creating an integrated process between a qualitative, social-science-based approach and a quantitative, natural-science-based approach. The European Environment Agency report 'Scenarios as tools for international environmental assessments' (Alcamo 2001) proposes a 'story-and-simulation (SAS)' approach to developing scenarios. A variant of the SAS model was used in this study, with the quantitative 'storyline' scenarios constituting the first step of the process.

REFERENCES

Alcamo, J. (2001). *Scenarios as Tools for International Environmental Assessments*. European Environment Agency, Copenhagen, p. 31.

Alcamo, J. and Nakicenovic, N. (1998). Long-term greenhouse gas emission scenarios and their driving forces. *Mitigation and Adoption Strategies for Global Change* **3**(2–4), 343–381.

Behrendt, H. (1996). Inventories of point and diffuse sources and estimated nutrient loads — a comparison of different river basins in Central Europe. *Water Science and Technology* **33**, 99–107.

Bhaskar, R. (1989). *Reclaiming Reality*. Verso, London.

Blondel, J. (1995). *Comparative Government*. Prentice-Hall, London.

Braybrooke, D. (1987). *Philosophy of Social Science*. Prentice-Hall, London.

Bremmer, I. and Taras, R. (eds) (1997). *New States, New Politics: Building the Post-Soviet Nations*. Cambridge University Press, Cambridge.

Committee on Global Change Research, B. o. S. D., Policy Division (1999). *Global Environmental Change*. National Research Council, Washington, DC.

Cosgrove, W.J. and Rijsberman, F. (2001). The making of the world water vision exercise. In: *World Water Scenarios*, F. Rijsberman (ed.). Earthscan, London.

Dawisha, K. and Parrott, B. (eds) (1997). *The Consolidation of Democracy in East-Central Europe*. Cambridge University Press, Cambridge.

EC (2000). *Directive 2000/60/EC of the European Parliament and of the Council of 23 October 2000* – establishing a framework for Community action in the field of water policy. *Official Journal of the European Communities*.

ENDS (1995). Strategic Rethink on Water Urged by MEPs.

European Communities (1997). *Agenda 2000: Commission Opinion on Estonia's Application for Membership of the European Union*. Office for Official Publications of the European Communities, Luxembourg.

Glenny, M. (1993). *The Rebirth of History*. Penguin, London.

Gooch, G.D. (2003). *Scenarios for Environmental Policy-Making on the Estonian–Russian Border*. Department of Management and Economics, Linköpings University, Linköping, p. 17.

Greeuw, S.C.H., van Asselt, M.B.A., *et al.* (2000). *Cloudy Crystal Balls*. European Environment Agency, Copenhagen.

Grimvall, A., Stålnacke, P. and Tonderski, A. (2000). Time scales of nutrient losses from land to sea – a European perspective. *Ecological Engineering* **14**, 363–371.

Held, D. (1987). *Models of Democracy*. Polity Press.

Hill, M. (1997). *The Policy Process in the Modern State*. Prentice-Hall, London.

Jordan, A. (2000). The politics of multilevel environmental governance: subsidiarity and environmental policy in the European Union. *Environment and Planning* **32**, 1307–1324.

Jordan, A.G. (1997). Overcoming the divide between comparative politics and international relations approaches to the EC: what role for 'post-decisional politics'? *West European Politics* **20**(4), 43–70.

Jordan, A.G. (1999). European Community water policy standards: locked in or watered down? *Journal of Common Market studies* **37**, 3–37.

Jouvenel, H.d. (2000). A brief methodological guide to scenario building. *Technological Forecasting and Social Change* **65**, 37–48.

Kahane, A. (1997). Imagining South Africa's future. In: *Learning from the Future*, L. Fahey and R.M. Randall (eds). Wiley.

Klijn, E.H. and Koppenjan, J.F.M. (1997). Beleidsnetwerken als theoretische benadering: een tus-senbalans. *Beleidsnetwerken* **2**, 143–167.

Kuhn, T.S. (1970). *The Structure of Scientific Revolutions*. The University of Chicago Press, Chicago.

La Spina, A. and Sciortino, G. (1993). Common agenda, southern rules: European integration and environmental change in the Mediterranean states. In: *European Integration and Environmental Policy*, J. Liefferink, P. Lowe and A. Mol (eds). Belhaven, London.

Losee, J. (1980). *A Historical Introduction to the Philosophy of Science*. Oxford University Press, Oxford.

Lääne, A., Pitkänen, H., Arheimer, B., Behrendt, H., Jarosinski, W., Lucane, S., Pachel, K., Ræike, A., Shekhovtsov, A., Svendsen, L. and Valotka, S. (2002). *Evaluation of the Implementation of the 1988 Ministerial Declaration Regarding Nutrient Load Reductions in the Baltic Sea Catchment Area*. Finnish Environment 524, Finnish Environment Institute, Helsinki, p. 195.

Marty, F. (1997). *International River Management – The Political Determinants of Success and Failure: Introduction and Research Design*. Universität Zürich, Zürich.
Melvin, N. (1995). *Russians Beyond Russia: The Politics of National Identity*. Royal Institute of International Affairs, London.
Mostert, E. (1998). River basin management in the European Union; how it is done and how it should be done. *European Water Management*, 26–35.
Mostert, E. (1999). Perspectives on river basin management. *Physics and Chemistry of the Earth Part B: Hydrology, Oceans and Atmosphere* **24**, 563–569.
Mourad, D.S.J., van der Perk, M., Gooch, G.D., Loigu, E., Piirimäe, K. and Stålnacke, P. (2005). GIS-based quantification of future nutrient loads into Lake Peipsi/Chudskoe using qualitative regional development scenarios. *Water Science and Technology* **51**(3–4), 355–363.
Muiznieks, N. (1997). Latvia: restoring a state, rebuilding a nation. In: *New States, New Politics: Building the Post-Soviet Nations*, R. Taras and I. Bremmer (eds). Cambridge University Press, Cambridge.
Nordlöf-Lagerkrantz, U. and Lagerkrantz, J. (1993). *Östeuropa*. C.E. Fritzes AB, Stockholm.
Outhwaite, W. (1987). *New Philosophies of Social Science*. St. Martin's Press, New York.
Petersson, O. (1995). *Hur styrs Europa? Den politiska maktutövning i Europas stater*. Fritzes förlag AB, Göteborg.
Petersson, O. (1998). *Kommunalpolitik*. Nordstedts Juridik AB, Stockholm.
Porter, M. (1985). *Competitive Advantage*. Free Press, New York.
Ringland, G. (2002). *Scenarios in Public Policy*. John Wiley & Sons, Chichester.
Sakwa, R. (2002). *Russian Politics and Society*. Routledge, London and New York.
Schultz, G.A. (2001). Integrated water resources management: the requirements of the European Union, the problems of environmental impact assessment, and implementation of the sustainable development principle. In: *Integrated Water Resources Management*. M.A. Marino and S.P. Simonovic (eds). IAHS Publication No. 272, pp. 3–12.
SOU (1998). *Regional frihet och statlig ansvar – en principiell diskussion*. Inrikes-departementet, Stockholm.
Stålnacke, P. (1996). *Nutrient loads to the Baltic Sea*. PhD Thesis, Linköping *Studies in Arts and Science* 146, 78 pp.
Stålnacke, P., Vandsemb, S.M., Vassiljev, A., Grimvall, A. and Jolankai, G. (2004). Changes in nutrient levels in some Eastern European rivers in response to large-scale changes in agriculture. *Water Science and Technology* **49**(3), 28–36.
Toresson, B. (1995). *Med Berlinmuren i backspegeln – på obanad väg mot framtiden*. PM Bäckström Förlag, Stockholm.
UN/ECE (1994). Protection and use of transboundary watercourses and international lakes in Europe. *Natural Resources Forum* **18**(3), 171–180.
Vagstad, N., Stålnacke, P., Andersen, H.E., Deelstra, J., Jansons, V., Kyllmar, K., Loigu, E., Rekolainen, S. and Tumas, R. (2004). Regional variations in diffuse nitrogen losses from agriculture in the Nordic and Baltic region. *Hydrology and Earth System Sciences* **8**(4), 651–662.
Ward, N., Lowe, P., *et al.* (1997). Implementing European water quality directives. In: *The Politics of Sustainable Development*, S. Baker, M. Kousis, D. Richardson and S. Young (eds). Routledge, London, pp. 198–216.
Whitehead, A.N. (1926). *Science and the Modern World*. Syndicate of the Press of the University of Cambridge, Cambridge.
Winch, P. (1958). *The Idea of Social Science*. Routledge.
Young, O.R. and Osherenko, G. (1993). Testing theories of regime formation: findings from a large collaborative research project. In: *Regime Theory and International Relations*, V. Rittberger and P. Mayer (eds). Clarendon Press, Oxford, pp. 223–251.

2

The physical environment: Lake Peipsi and its drainage basin

P. Stålnacke, S. Vandsemb, T. Nõges, P. Nõges, D. Mourad, M. van der Perk, S. Nilsson, F. Hannerz, S. Langaas, E. Loigu, A. Heinsalu and B. Skakalsky

2.1 INTRODUCTION

Lake Peipsi (Chudskoe in Russian) (Figure 2.1) is one of the largest European lakes with a surface area of 3500 km^2. Due to its shallow character (its depth does not exceed 15 m) together with anthropogenic activities in the river basin such as agriculture and settlements, it is sensitive to eutrophication problems (Nõges 1996) such as water blooms in summer and autumn caused by blue-green algae (cyanobacteria), and often accompanied by summer fish-kills. This is further assessed in Section 2.4.

The drainage basin is part of the Baltic Sea drainage basin. Lake Peipsi discharges into the Narva River, which in turn has its outflow in the Gulf of Finland (Figure 2.1). In this outline of geographical features, we will hereafter focus upon the Lake Peipsi drainage basin and disregard the Narva River basin.

 Integrated Transboundary Water Management in Theory and Practice: Experiences from the New EU Eastern Borders edited by Geoffrey D. Gooch and Per Stålnacke. ISBN: 1843390841. Published by IWA Publishing, London, UK.

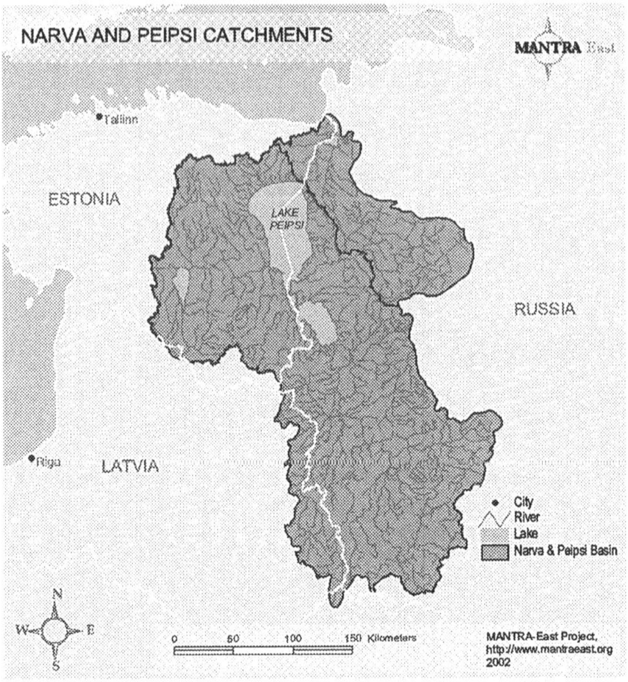

Figure 2.1 The Narva River/Lake Peipsi and its drainage basin.

This chapter summarises the major physical features of the region, including:

- The geographical features of Lake Peipsi and its drainage basin.
- Assessments on the emissions and flows of nutrients from the drainage areas of Lake Peipsi, with main emphasis on the nutrients' effects.
- Assessments of the water quality and ecological status of the lake.

The results from these assessments will be used as input to the methodological processes of translating the qualitative scenarios to quantitative ones (Chapter 5). This chapter should also be seen in relation to the social-economic and politico-administrative features of the region that is given in the next chapter (Chapter 3).

2.2 GEOGRAPHICAL FEATURES OF THE LAKE PEIPSI AND ITS DRAINAGE BASIN

The Lake Peipsi drainage basin (Figure 2.1) has a size of almost 45,000 km^2 and is thus approximately 12 times larger than the surface of the lake. The basin is shared by Russia (59%), Estonia (33%), Latvia (8%) and Belarus (0.3%). The largest sub-catchment is the Velikaya River basin (located in Russia), draining approximately 58% (25,800 km^2 mean discharge: 195 m^3/s) of the whole Lake Peipsi drainage basin. The Emajõgi River basin (in Estonia) is the second largest sub-catchment, covering approximately 20% (8750 km^2 mean discharge: 68 m^3/s)

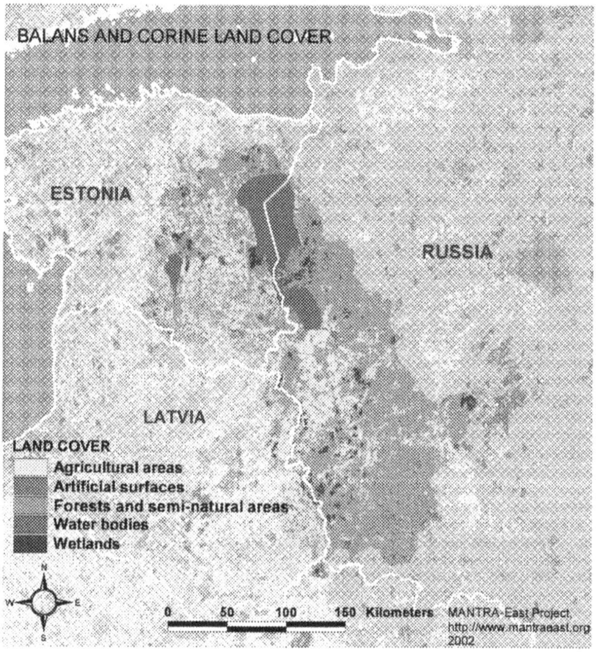

Figure 2.2 Land cover in the Lake Peipsi drainage basin.

of the total basin. These two rivers account for nearly 65% of the water discharge into the lake (Stålnacke *et al.* 2002), and thus are the most important when it comes to nutrient input into the lake.

The dominating land cover class is forests and semi-natural areas, covering approximately 63% of the basin (Figure 2.2). For clarification, semi-natural areas comprise: (i) patchy areas with a mixture of agricultural and forest, (ii) shrublands and (iii) non-agricultural open areas. Agricultural land covers around 30% of the basin. This is found mainly in Latvia and Estonia, while forests and semi-natural areas constitute most of the Russian part of the basin. Wetlands are scattered throughout the basin, although a relatively large portion are found near the shores of Lake Peipsi.

The population in the basin is approximately 1.1 million people. The two largest cities, Pskov (Russia) and Tartu (Estonia), with 206,000 and 120,000 inhabitants, respectively, are the two largest point sources of nutrients and pollution in the drainage area (further details about point sources are given in Section 2.3).

The drainage basin is flat, with a highest point of only 296 m above sea level and an average elevation of 163 m. This basin is a typical North-European lowland area of glacial origin, characterised by Palaeozoic bedrock, covered by unconsolidated glacial materials of variable thickness.

According to the revised FAO soil-coding system (FAO 2000), the Russian part of the Lake Peipsi drainage basin is dominated by podzoluvisols and gleysols, while luvisols are abundant in the Latvian and Estonian part of the basin (Figure 2.3).

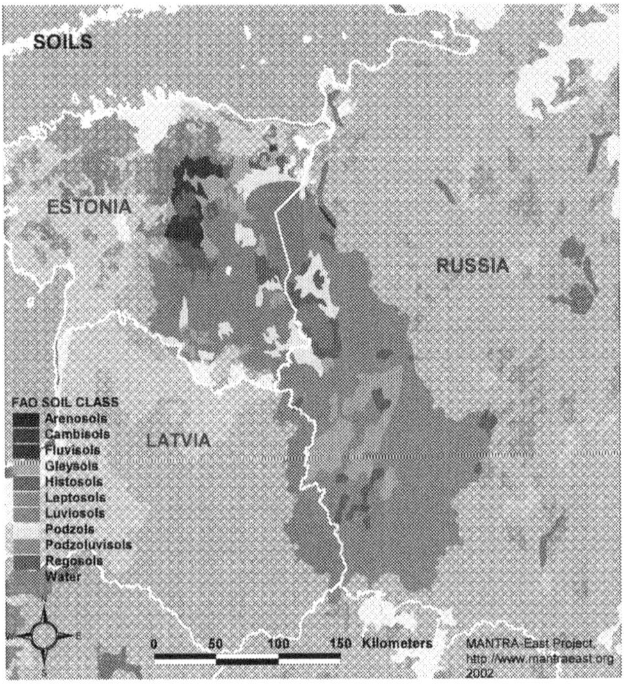

Figure 2.3 Soil classes in the Lake Peipsi drainage basin according to the revised FAO soil-coding system.

The lake consists of three parts (Figure 2.4). The area of the largest northern part, Lake Peipsi *sensu stricto* (Lake Peipsi s.s.) is 2611 km^2, with a mean depth 8.3 m and the greatest depth 12.9 m. The southern part, Lake Pihkva (Pskovskoe in Russian) with an area of 708 km^2 and a mean depth of 3.8 m is connected with Lake Peipsi by the narrow river-shaped Lake Lämmijärv having an area of 236 km^2 and a mean depth of 2.5 m. The Lake Peipsi s.s, is an unstratified eutrophic lake with some mesotrophic features, Lake Lämmijärv has some dyseutrophic features, while Lake Pihkva is an unstratified eutrophic, or even hypertrophic water body.

The amount of water in Lake Peipsi is 25 km^3 and the residence time of the water is about 2 years. The outflow is through the River Narva, which flows into the Gulf of Finland (Jaani 2001). The water of Lake Peipsi is normally wind-mixed down to the bottom during the ice-free period. During this time there is no stratification, the whole water mass is well aerated, and the surface layer of bottom sediments is aerobic. The ice-cover lasts from November to December until April isolating the water from the atmosphere and wind action. At the end of severe winters as, for example in 2003, oxygen depletion may occur in the deeper water layers and bottom sediments may get anaerobic, first of all at the deepest 15.6-m site in Lake Lämmijärv. It may cause a release of iron-bound phosphorus from the sediments. No winter oxygen depletion has been reported in Lake Peipsi s.s. and fishes are safe there in winter. In the shallow and more eutrophic Lake Pihkva, the risk of winter oxygen depletion is quite realistic, especially in long

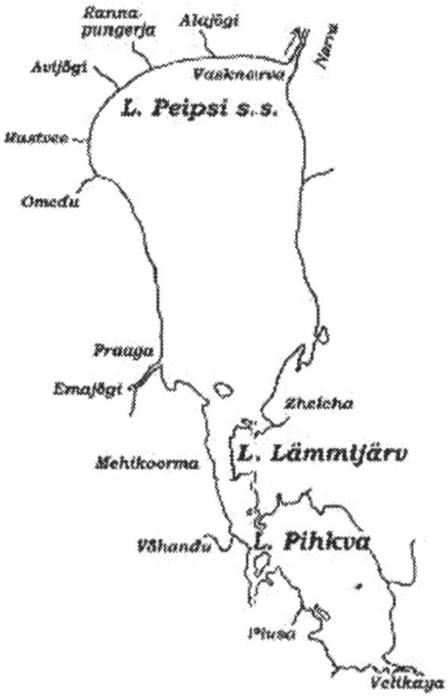

Figure 2.4 The three parts of Lake Peipsi and the major rivers.

and cold winters coinciding with low water level when the under-ice storage of oxygen is small.

Lake Peipsi is important for recreation and fishery, and taking into account the lakes annual fish catches (9000–12,000 tons or 25–34 kg/ha) Lake Peipsi is the most important in this respect of all the large lakes in Northern Europe. The main commercial fishes are lake smelt (*Osmerus eperlanus eperlanus*), perch (*Perca fluviatilis*), pike-perch (*Sander lucioperca*) and bream (*Abramis brama*). Up until the 1990s vendace (*Coregonus albula*) were also important, but the stock of this formerly important commercial fish decreased sharply in the beginning of 1990s and has not recovered. This decrease can be explained by the occurrence of unfavourable spawning conditions in several successive years, as well as the impact of eutrophication. At the same time, the abundance of pike-perch increased remarkably (Pihu and Kangur 2001). The benthic fauna of Lake Peipsi has been subjected to strong modification by invasion of zebra mussels in 1930s, and a gammaridean amphipod from Lake Baikal in the 1970s. Nevertheless, high species diversity, stable abundance of the benthic community and survival of sensitive clean-water species demonstrates reasonably good water quality (Timm *et al.* 2001).

2.3 POINT AND NON-POINT SOURCE POLLUTION IN THE LAKE PEIPSI BASIN

2.3.1 Point sources

Nutrients from point sources are discharged into the rivers of the Lake Peipsi basin mainly via discharges from wastewater treatment plants (WWTPs) and oil-shale

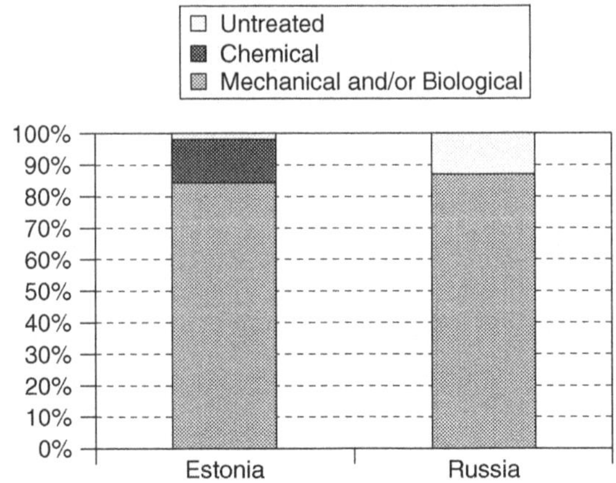

Figure 2.5 Share of different wastewater treatment technologies in Estonian and Russian part of Lake Peipsi basin in 2001, respectively.

mines. As part of the methodology, pollution load data was collected to assess and evaluate possible changes during the last decade of the 20th century. The assessment was based on national statistics kept by the Data and Information Centre of the Estonian Ministry of Environment, Neva-Ladoga Basin Water Management Administration and Pskov Regional Committee for Environmental Protection. Priority for the assessment of point source was given to the estimation of emissions from the largest cities and sewage treatment plants (Pskov, Tartu, Ostrov, Opochka, Gdov, Novoizborsk, Pitalovo, etc.), and settlements along the shoreline of the Lake Peipsi–Pihkva. Calculations were primarily based on monitoring data supplemented with data on the number of people connected to wastewater systems, combined with assignment of emission coefficients (e.g. per capita emissions) and data on treatment efficiency.

Both industrial and municipal effluents are nowadays generally connected to WWTPs (98% in Estonia and 87% in Russia; Figure 2.5). In Estonia, the share of biological treatment with phosphorus removal is still at a modest level and mechanical treatment still dominates. Tertiary (chemical) treatment is rare in both countries (Figure 2.5). Noteworthy is also the decrease in point source pollution of nutrients and organic matter throughout the 1990s in both countries (Figure 2.6).

The largest point pollution source in the drainage area is Pskov City with its ca. 200,000 inhabitants that emits its treated wastewaters directly into the Velikaya River. In the Russian part of the basin, the major single point sources are: ME Pskov GORVODOKANAL, Slantsy town slate-extracting enterprises – OJSC LENINGRADSLANETS, OJSC Slantsy Factory, OJSC TSESLA Slantsy Cement Plant, ME Ostrov VODOKANAL, ME Ivangorod HCS, LLC Pechory VODOKANAL and MUE Slantsy HCS. Their total share comprises over 88% of the total wastewater discharge from the Russian part of the basin. A still major unsolved issue is the contribution of nutrients from point sources in Russia and especially from the WWTP in Pskov City. According to recent official data obtained from the joint Estonian–Russian Transboundary Water Commission the emissions from point source load to the River Velikaya were 63 and 445 ton for phosphorus and nitrogen, respectively.

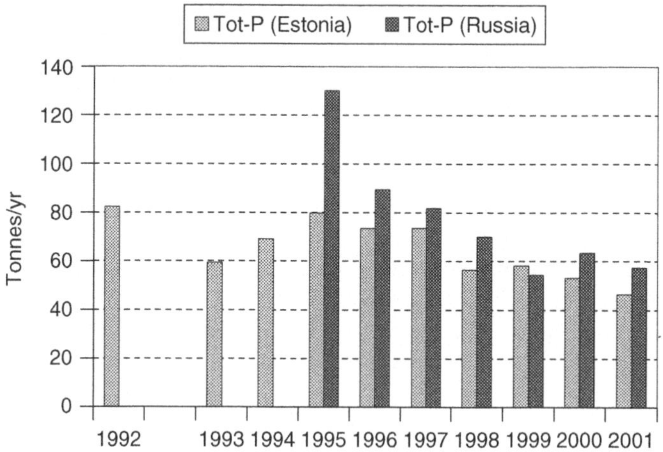

Figure 2.6 Point source emission from WWTPs in the Lake Peipsi drainage basin. Total-P in Russia estimated from inorganic-P according to figures from St. Petersburg WWTP (Skakalsky, personal communication).

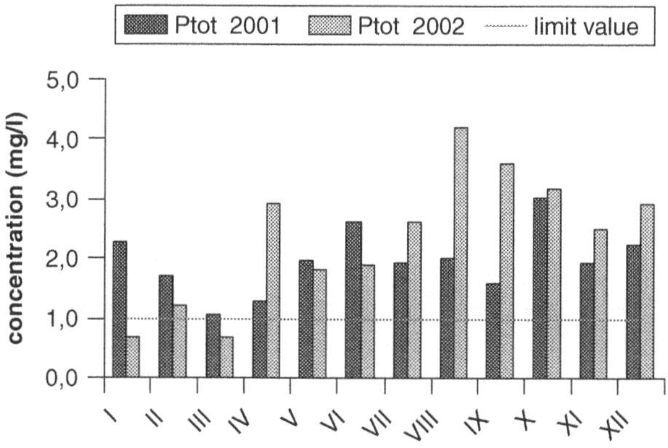

Figure 2.7 Monthly effluent concentration of P from Tartu WWTP in 2001 and 2002. The limit value indicates the standard maximum concentration permitted.

If we assume that there are 200,000 inhabitants in Pskov City, they produce according to well-established standard coefficients 200 ton P per year (3 g P/day capita). This means that the Pskov WWTP has a treatment capacity of nearly 60%, which is really high given that only mechanical and biological treatment is in operation. According to normal standards a WWTP with mechanical and biological treatment normally has a treatment efficiency of maximum 30–35% for phosphorus.

The second largest point source in the basin is Tartu City in Estonia (120,000 inhabitants) that emits its sewage water to the Emajõgi River. Despite increased sewage treatment with a P removal of approximately 80%, the wastewater concentration emitted after treatment was higher than the maximum permitted standard of 1.0 mg/l (Figure 2.7). In fact, high level of phosphorus emissions can be observed in basically all small towns and villages in Estonia. The reason for this is the relative high pollution charge that was implemented by the authorities that led to a dramatic decline in the water consumption both by industries and private households. This can, for example, be seen in the figures for the per capita water consumption that decreased from around 200–250 l to as low as 70–80 l per capita. Paradoxically, these figures indicate on the one hand sustainable use and management of water resources, but on the other hand led to too high inlet concentrations of P and N to the WWTPs that despite modern technology causes problem in the treatment processes.

Oil-shale mining waters form the other main source of point pollution, which is mainly treated in mechanical sedimentation ponds. Mining waters do not have a significant impact on organic and nutrient pollution in an overall Lake Peipsi perspective (Stålnacke *et al.* 2005). Nonetheless, mining waters are impure, containing considerable amounts of pollutants such as phenols and even hydrocarbons.

2.3.2 Diffuse pollution sources and land-use changes

It is now well known that the dynamics of diffuse nutrient emissions in large catchments is complicated, and entails significant temporal and spatial variability. This makes it difficult to empirically derive simple cause and effect relationships. In addition, it is also widely accepted that nutrient concentrations which originate from diffuse sources depend on factors such as the following: flow conditions; hydrological pathways; the occurrence of other nutrients or micronutrients; bioavailability; processes such as mineralisation, immobilisation and denitrification; and the ratio between N and P. However, little is yet known about the relative importance of, and the interactions between, these factors in time and space. It has already been pointed out in several articles (e.g. Hirsch and Slack 1984) that prevailing weather conditions often cause natural fluctuations in time series of nutrient concentration values, and this may impede the detection of existing trends. Nutrient losses are also affected by changes in crop yields, which, in turn, are largely influenced by the weather conditions. Current hydrometeorological conditions other than water discharge can have an impact on phosphorus losses. For example, Vassiljev (1994) conducted a study in Estonia and found that 20 times more phosphate phosphorus was released from frozen soils than from thawed soils. Accordingly, during periods that include several mild winters, it can be expected that losses will either increase due to an elevated risk of soil erosion, or decrease due to a drop in the freezing out levels of plant residues, especially on grassland. In both cases, a detected trend might be falsely interpreted as being the result of changes in agricultural practices.

In the late 1980s, the use of commercial fertilisers in most Eastern European countries began to decrease at an unprecedented rate (Table 2.1). For example, in Estonia and Latvia, sales of mineral fertilisers dropped by a factor of 15 from 1987 to 1996, and the number of livestock decreased by a factor of almost 4 during the same period (Vagstad *et al.* 2004). Moreover, the import of feedstuff in Eastern Europe became almost zero, and extensive slaughtering of livestock reduced the amount of manure (Oláh and Olàh 1996; Tonderski 1997; Löfgren *et al.* 1999; Vagstad *et al.* 2004). Such abrupt and substantial changes in land use have

Table 2.1 Total consumption of commercial fertilisers (kg tonnes) in Eastern European countries from 1987 to 1998.

Year	1987	1988	1989	1990	1991	1992	1993	1994	1995	1996	1997	1998
Albania	95	97	107	102	31	24	18	15	11	7	5	23
Bulgaria	745	903	807	685	472	237	205	225	123	192	200	169
Croatia						184	139	191	184	202	276	177
Czech Republic							256	337	348	347	314	300
Czechoslovakia	1556	1600	1561	1304	425	390						
Estonia	265	244	238	198	195	169	65	41	27	22	29	35
Hungary	1373	1418	1206	680	326	189	293	314	368	455	427	389
Latvia	525	501	448	382	359	169	86	54	19	26	34	33
Poland	3277	3415	3310	1541	1122	1192	1282	1427	1511	1596	1700	1557
Romania	1390	1355	1378	1213	463	553	602	374	363	422	315	360
Slovakia							81	99	108	119	107	119
Slovenia						63	54	86	74	63	74	77
Yugoslavia SFR	1031	1015	855	828	581							
Yugoslavia, Federal Republic						99	72	84	143	215	244	202

Sources: FAO (2002), Statistics Estonia and Statistics Latvia.

rarely been recorded in the history of modern European agriculture, and they have created unique opportunities to study the impact of agricultural practices on water quality. This is further assessed in Section 2.4.

The collapse of the Soviet Union in 1991, and the simultaneous conclusion of its power in Central and Eastern Europe, brought massive changes to the region's agriculture. In communist times, the land use in the north-western part of the Soviet Union had always been more intensive than in other regions. The collective and state farms were predominantly specialised in meat and dairy production, and additional growing of fodder crops. Due to inefficient use of fertiliser and the lack of proper manure handling, agricultural methods resulted in high emissions of nutrients N and P into rivers and lakes, and subsequent eutrophication problems (Loigu and Leisk 1996). After the 1991 revolution, the Baltic States became independent, and a coincidence of developments caused the stagnation of agricultural economy in the Baltic States. Unwin (1997) gives a total overview of factors, of which the most important arc:

- The collapse of the traditionally important Russian market, especially after the introduction of import tariffs in 1994.
- The impossibility to penetrate the European Union (EU) market due to its protectionist policies, while in the mean time import of cheap EU products is not hindered.
- The general confusion about privatisation and land reform, in which a coherent national policy is absent. In the Russian part of the river basin, agricultural economy stagnated as well, mainly as a result of the deep economic crisis in Russia.

At present, the area has already been in a period of transition for more than a decade, and the future is still highly uncertain. Although economy will grow definitely, especially on the western side of this new EU border, it is not yet clear whether the environment will gain or loose from that. The decline of agriculture during the 1990s actually caused diffuse pollution to decrease and the quality of rivers to rise. So, in the future, it is possible that water quality will decrease again, if agriculture recovers. On the other hand, water quality may benefit from EU regulation and good transboundary cooperation to agree on farm mineral balances and better public wastewater treatment. Thus, nutrient emissions and water quality are linked to future economic development in various, sometimes indirect ways.

Results from modelling the diffuse nutrient sources are given in Chapter 4.

2.4 RIVER NUTRIENT LOADS IN THE DRAINAGE BASIN

Increases in water discharges usually result in subsequent increases in nutrient loads. Also, as hydrological processes and pathways determine the mobilisation and transport of nutrients, nutrient fluxes through rivers are related to the hydrological behaviour of the area and the resulting river discharge. Due to the Nordic climate, characterised by cold winters in which much precipitation is stored in the

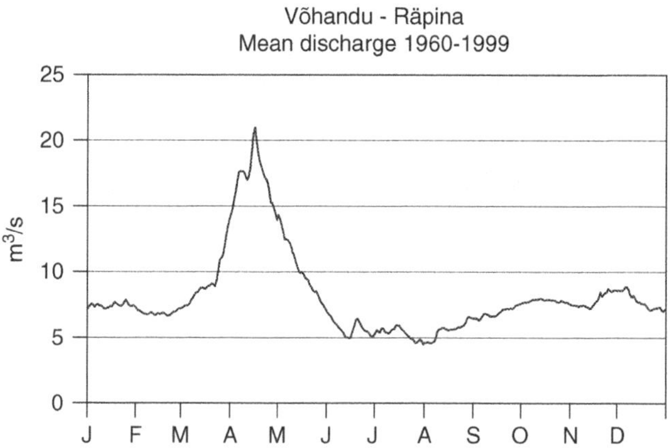

Figure 2.8 Mean annual hydrograph (period: 1960–1999) of the Võhandu River at Räpina (Estonia).

form of snow, all rivers in the area show a similar, typical pattern of discharge throughout the year, which can be illustrated by mean of the annual hydrograph of the Võhandu River at Räpina (Figure 2.8). Distinct seasonal changes are visible in the hydrograph which are characterised by a pronounced springtime peak in which discharge is up to four times higher than base flow level. This springtime peak is generally the result of large amounts of surface runoff from melting snow or rainfall, which cannot infiltrate into the still frozen, and thus impermeable, soil. A secondary seasonal increase in discharge can be seen in the autumn, which is caused by increased precipitation in that season, combined with a diminished uptake of water by vegetation. During the rest of the year, base flow level is only increased during short periods as a result of rainstorms and rainy days.

As nutrient loads are the product of concentrations and discharges over time, the characteristics of water flow change shown in the hydrograph are important for the yearly load of nutrients discharged by the rivers. During base flow conditions, and periods of high biological uptake of nutrients during the summer, discharges as well as nutrient concentrations are generally low, while both discharge levels and concentrations are significantly higher during the springtime. This means that the spring season is the most crucial period for nutrient export. For a correct assessment of total nutrient loads into the lake, river discharge and nutrient concentrations need to be available for all periods of the year, based on a sufficient sampling frequency. At least for the Estonian part of the drainage basin, regular river data are available for the past decades, including concentrations of major nutrient ions and total nitrogen (TN) and total phosphorus (TP) (both in mg/l), and discharge (in m^3/s). Unfortunately, data for the Russian side of the lake are largely unavailable and the number of sampling sites is substantially lower. Such 'unbalanced' data is rather typical for transboundary basins and illustrate the critical issue of a sound harmonised assessment of basic information such as riverine load. Nonetheless, a recent joint Estonian and Russian expedition in August 2003

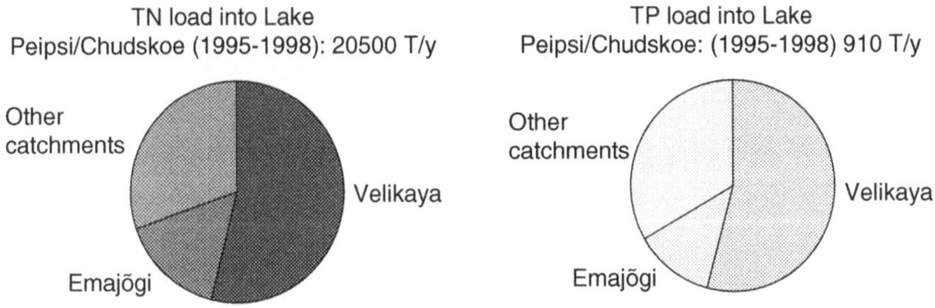

Figure 2.9 Total loads into Lake Peipsi/Chudskoe and the relative contribution of the two major rivers (according to Stålnacke *et al.* 2002).

revealed extremely low N:P ratios (less than 8:1) in the mouth area of the River Velikaya and generally in Lake Pihkva (Külli Kangur, personal communication). These results without any doubt confirm the high P load and impact from the River Velikaya on Lake Pihkva.

Stålnacke *et al.* (2002) estimated the total loads into the lake by a methodology based on the extrapolation of loads for rivers with lack of data. The results indicate mean annual loads for the 1995–1998 period (Figure 2.9). The Emajõgi River delivers 3238 ton/year of TN, while the Velikaya River was estimated to discharge about 11,050 ton/year of TN into the lake. This represents about 70% of the total load into the lake, which was estimated to be 20,500 ton/year of TN. For TP, they estimated that the River Emajõgi delivers 113.5 ton/year into the lake, and the Velikaya River even 493 ton/year. This represents about 80% of the total estimated mean annual TP load into the lake.

A number of studies (e.g. Stålnacke and Grimvall 2000) have shown that the level of nutrient losses and riverine loads during individual years is influenced primarily by variations in precipitation and water discharge. This is also the case in the Lake Peipsi basin as illustrated by Figure 2.10. More precisely, it can be noted that the load is for most years correlated with the water discharge.

Despite this, the data from the Emajõgi River (Figure 2.10) also show a rather clearly downward trend in the TN loads. For TP, the trends are less clear or even absent. The influence of the mean annual discharge is clear. It is also clear that the higher loads in the second half of the 1980s were exaggerated by the rather high discharges during that period, and that the dry year of 1996 generally delivers the lowest load in the entire period for all rivers.

A more thorough investigation of trends in Estonian streams and rivers detected 20 statistically significant downward trends in TN (one-sided test at the 5% level) from a total of 22 sites (Table 2.2; Iital *et al.* 2005). The study represented both small and large sub-catchments of the entire Estonian part of the Lake Peipsi drainage area (100–48,000 km^2).

These observed decreases in TN were by the authors explained by (i) substantial reductions in the use of organic and inorganic fertilisers, (ii) lessening of cultivated and ploughed areas, and increased proportions of grassland and abandoned land and (iii) improvements in farm management practices. Insufficient maintenance

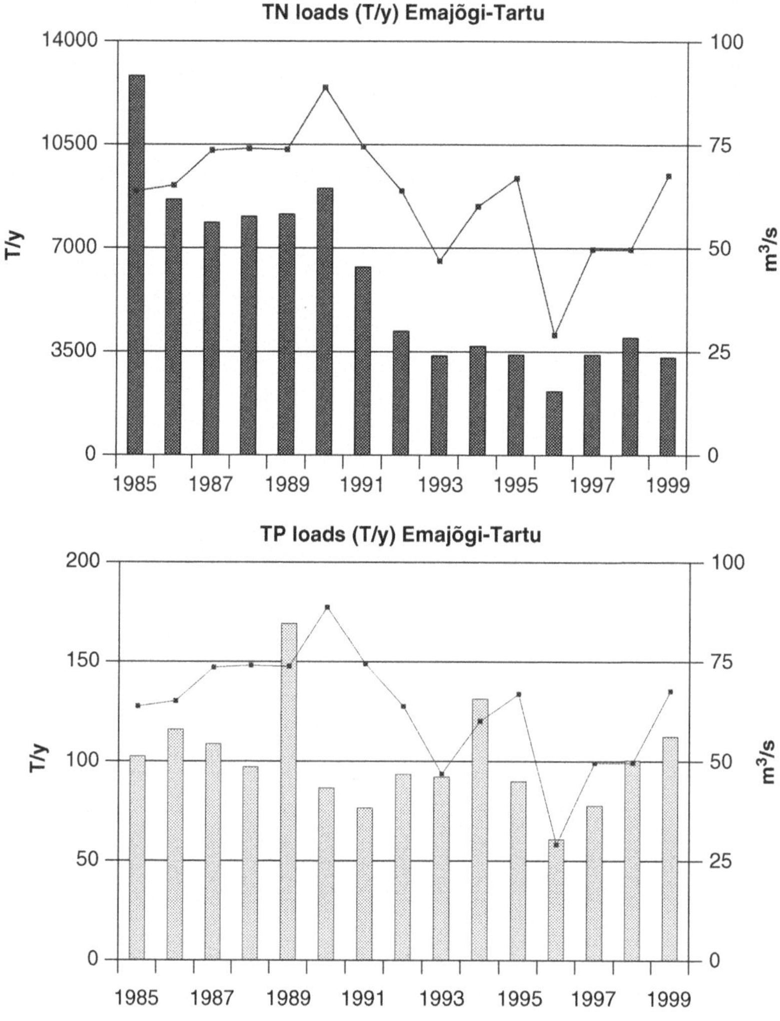

Figure 2.10 Illustration of the interannual variability in nutrient loads (the bars) and water discharge (the line) in a typical river in Lake Peipsi basin. Concentration data obtained from Estonian Environment Information Centre.

of main ditches during the past decade, resulting in overgrowth of bushes and macrophytes, which in turn enhanced the retention potential (i.e. denitrification and biological uptake) was also pointed out as an additional explanation. In Estonia, trend studies of this kind have previously been performed chiefly in single, smaller catchments. Loigu and Vassiljev (1997) noted a clear downward trend in nitrate concentrations in the rivers of the small and intensively cultivated Kurna catchment (23.2 km^2, 68% arable land) during the period 1987–1996. This trend was particularly conspicuous for the peak concentrations, and the cited authors stated that it could be explained by the decreased application of fertilisers and agricultural

Table 2.2 Statistical trend test on the long-term trends for nutrient concentrations in Estonian rivers. After Iital *et al.* (2005). PMK refers to a recently modified version of the seasonal Mann–Kendall test (Libiseller and Grimvall 2002), which has been adapted to account for the influence of confounding (i.e. meteorological or hydrological) variables; in this case water discharge as such a variable. Statistically significant trends marked in bold font ($P < 0.05$).

Sampling site	Total N			Total P		
	Years monitored	PMK-Stat	*P*-value	Years monitored	PMK-Stat	*P*-value
Alajõgi-Alajõe	1984–2001	−1.85	**0.032**	1984–2001	2.20	**0.014**
Ahja-Kiidjärve	1985–2001	−3.85	**<0.001**	1981–2001	−1.20	0.114
Põltsamaa-Rutikvere	1986–2001	−3.71	**<0.001**	1986–2001	−1.34	0.091
Rannapungerja-Roostoja	1992–2001	0.06	0.475	1992–2001	−1.41	0.079
Emajõgi-Jõesuu	1987–2001	−3.43	**<0.001**	1987–2001	0.87	0.193
Võhandu-Räpina	1986–2001	−3.92	**<0.001**	1986–2001	−0.74	0.231
Võhandu-Himmiste	1991–2001	−2.76	**0.003**	1986–2001	−1.30	0.098
Tänassilma-Oiu	1986–2001	−3.59	**<0.001**	1986–2001	−3.75	**<0.001**
Piusa-Korela	1992–2001	−2.35	**0.009**	1992–2001	−0.29	0.386
Avijõgi-Mulgi	1992–2001	−2.25	**0.012**	1992–2001	−1.31	0.094
Tagajõgi-Tudulinna	1992–2001	−2.26	**0.012**	1992–2001	0.27	0.393
Kääpa-Kose	1986–2001	−2.30	**0.011**	1986–2001	−0.07	0.472
Tarvastu-Põdraoja	1986–2001	−3.35	**<0.001**	1986–2001	−2.61	**0.005**
Pedja-Jõgeva	1986–2001	−0.01	0.495	1987–2001	−0.83	0.202
Pedja-Tõrve	1986–2001	−2.51	**0.006**	1987–2001	−0.51	0.304
Väike-Emajõgi-Pikasilla	1986–2001	−3.79	**<0.001**	1986–2001	−0.27	0.393
Emajõgi-Tartu	1984–2001	−2.06	**0.020**	1982–2001	1.02	0.154
Emajõgi-Kavastu	1986–2001	−1.75	**0.040**	1986–2001	−0.86	0.196
Narva-Vasknarva	1992–2001	−1.87	**0.031**	1992–2001	1.05	0.146
Õhne-Suislepa	1986–2001	−3.58	**<0.001**	1986–2001	2.31	**0.010**
Õhne-Roobe	1991–2001	−2.36	**0.009**	1986–2001	1.29	0.099
Porijõgi-Reola	1992–2001	−2.99	**0.001**	1992–2001	0.04	0.485

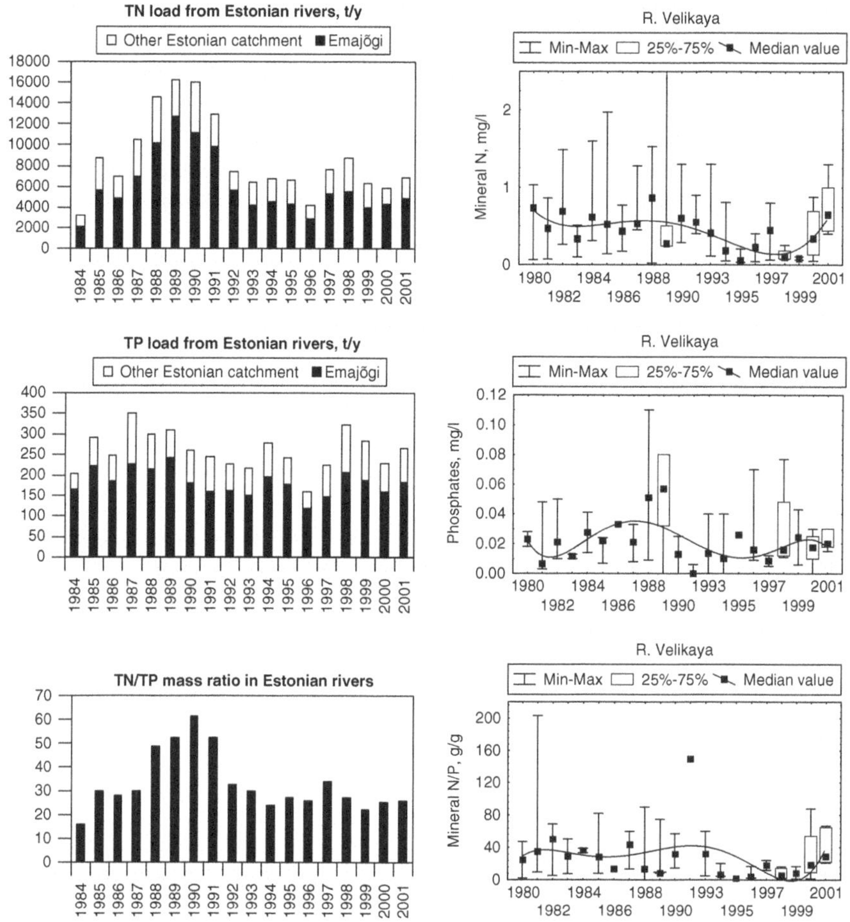

Figure 2.11 Annual riverine loads of TN and TP and their ratio in the Estonian rivers of Lake Peipsi drainage basin (left panels), and concentrations and ratio of mineral N and P in the River Velikaya (Russia) (right panels). *Source:* After Nõges *et al.* (2005).

production in the catchment. Similarly, Mander *et al.* (2000) found downward nitrogen and phosphorus trends in another small (258 km^2) agricultural catchment in Estonia. However, for TP, significant downward trends for TP were by Iital *et al.* (2005) detected only at two sampling sites, and there were also upward trends at two sites (Table 2.2; Iital *et al.* 2005).

Similar conclusions where also drawn by Nõges *et al.* (2003) in their estimation of the riverine loads from Estonian rivers (Figure 2.11). The same study also attempted to assess the trends in the Russian part of the basin. The results indicate that, in the largest Russian river, Velikaya, phosphorus concentration first decreased sharply at the beginning of the 1990s, but has then remained at the same (or even higher) level as in the mid-1980s. The nitrogen concentrations (as inorganic N) show the same tendency as in the Estonian ones with a rather sharp decline until the mid-1990s. The increased tendency in N concentrations in the late 1990s and early 2000s is though difficult to explain (Figure 2.11)

Both the studies by Nõges *et al.* (2003) and Iital *et al.* (2005) with different trends for nitrogen and phosphorus also mean a significant decrease in the N/P ratio. More precisely, the N/P-ratio decreased from around 80–100 in 1987 to 12–15 in 1991 (Iital *et al.* 2005). Low N/P ratios can increase the probability that nitrogen will limit primary production in the lake instead of phosphorus and thus enhance the risk of cyanobacterial blooms (see Section 2.5).

2.5 WATER QUALITY OF THE LAKE

If the water quality criteria developed for Estonian small lakes (Ott 2001) is applied to the entire Lake Peipsi, the lake falls into the 'good' class according to chlorophyll and phosphorus concentrations (Table 2.3) while nitrogen concentration and water transparency describe the lake as holding a rather 'moderate' status (Figure 2.12).

There have been quite small changes among dominant species and abundance of phytoplankton during the last 40 years. Observing the changes in blue-green algal dominance since 1962, one can notice its increase in Lake Peipsi s.s. in the late 1960s and 1970s, a decline in 1980s and a new increase since the 1990s and up to the present (Figure 2.13). Cyanobacterial blooms are mainly attributed to declining

Table 2.3 Indices reflecting trophic status of Lake Peipsi (Laugaste *et al.* 2001; Starast *et al.* 2001).

Parameter	Unit	Years	Mean	Lake Pihkva	Lake Lämmijärv	Lake Peipsi s.s.
TP	μg/l	1985–1996	42	63	53	35
TN	μg/l	1985–1996	768	1010	923	678
Chlorophyll a	μg/l	1983–1997	17.3	26	25	14
Secchi depth	m	1983–2000	1.76	1.25	1.42	1.8

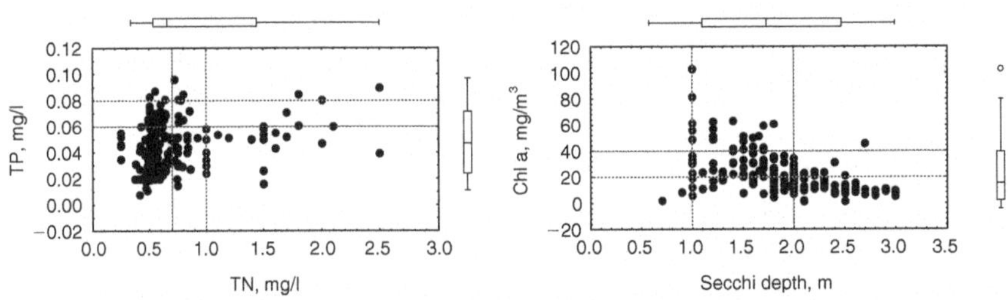

Figure 2.12 Concentration of chlorophyll a (Chl a), TP, TN and Secchi depth in Lake Peipsi in 1993–2001. The bands between dashed lines indicate 'moderate' status according to Ott (2001). Median, 10- and 90-percentiles, and non-outlier range are shown as box and whiskers.

water quality. In case of Lake Peipsi cyanobacteria constituted about one-half of the summer phytoplankton biomass already 40 years ago, and their blooms have been documented since the 19th century (Laugaste *et al.* 2001). As a paradox, the blooms almost disappeared in the conditions of heavy nitrogen loading in the 1980s and re-emerged in the 1990s when the nitrogen loading was reduced.

On the status and changes of other components of the ecosystem the following can be noted by Nõges *et al.* (2003a): In the period 1965–1999, the number of, but not the total biomass of crustaceans increased as a result of an increase in the abundance of small-sized zooplanktons (*Chydorus sphaericus*, *Bosmina longirostris*, *Mesocyclops oithonoides*) and a decrease in the abundance of larger species (*Bosmina berolinensis*, *Daphnia galeata*, *Leptodora kindti*, *Bythotrephes longimanus*, *Eudiaptomus gracilis*). The individual mean weight of zooplankton is often used as an indicator to characterise trophic relations, as it is a result of fish pressure on zooplankton, and reflects the trophic status of the waterbody (Haberman 1997; 2002). Eutrophication is generally accompanied by a decrease in the mean individual weight of zooplanktons. In Lake Peipsi the mean individual weight of zooplankton has decreased from 6.0 to 4.4 μg WW/individual since 1965. Slight changes in zooplankton species composition and mean weight do not however necessarily result in undesirable disturbances to the balance of organisms present in Lake Peipsi and the ecological status of the lake can be evaluated as 'good'.

Based on aquatic macrophytes, the ecological quality of Lake Peipsi can be estimated partly as 'good' and partly as 'moderate' depending on the region of the lake and on the exposure to wind and waves (Nõges *et al.* 2003a).

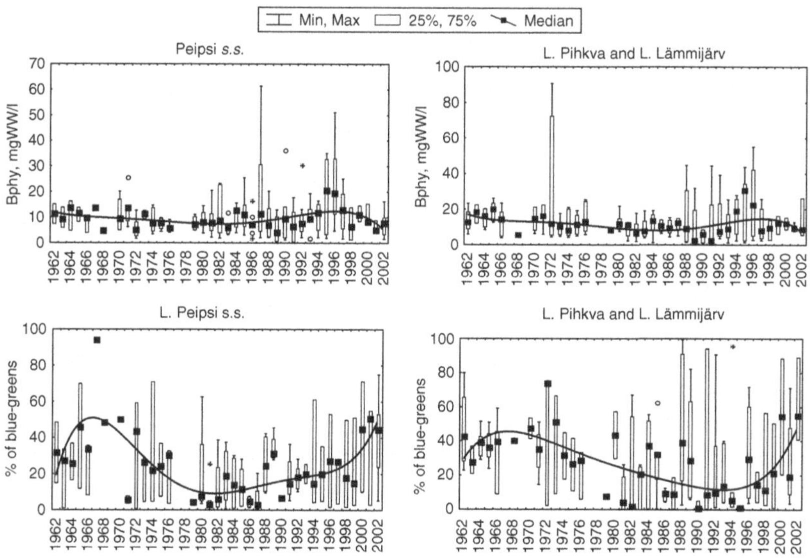

Figure 2.13 Long-term changes of phytoplankton biomass and blue-green dominance in Lake Peipsi s.s., and lakes Pihkva and Lammijärv. Data from May, July and October collected by Vörtsjärv Limnological Station, by Pskov Pedagogical Institute and by the Hydrometeorological Survey (from Nõges *et al.* 2005).

According to the fact that the benthic fauna of Lake Peipsi has been strongly modified by invasion of zebra mussel in 1930s, and a gammaridean amphipod from Lake Baikal in the 1970s (Timm *et al.* 2001), the ecological quality has to be evaluated as 'moderate' with respect to reference conditions. Nevertheless, high species diversity, and the survival of sensitive clean-water species and their stable abundance demonstrate 'good' water quality (Nõges *et al.* 2003a).

The main commercial fishes in Lake Peipsi are lake smelt (*Osmerus eperlanus eperlanus*), perch (*Perca fluviatilis*), pike-perch (*Sander lucioperca*), bream (*Abramis brama*), and until the 1990s also vendace (*Coregonus albula*). However, the stock of the formerly important commercial fish vendace decreased sharply at the beginning of the 1990s and has not yet recovered. This has been explained by the occurrence of non-favourable spawning conditions during several successive years, but the impact of eutrophication cannot be neglected. At the same time, the abundance of pike-perch increased remarkably (Pihu and Kangur 2001). The fish stock in Lake Peipsi is heavily exploited and over fishing was common already in the 19th century. As sharp decreases of clean-water species like vendace and whitefish, episodic fish-kills, reduction of stocks and the abundance of older age classes of top predators like pike-perch, and increases of omnivores and habitat generalists like ruffe the ecological quality cannot be evaluated as higher than 'moderate' (Nõges *et al.* 2003a).

2.5.1 The history of the lake: according to diatom analyses

It is also possible to reconstruct the long history of a lake by analysing remains of plants and animals preserved and accumulated in lake sediments throughout its existence. Among biological remains preserved in lake sediment, diatoms are particularly good indicators of environmental change because they are abundant in all aquatic environments and are highly sensitive to water quality changes (Stoermer and Smol 1999). They possess very rapid replication and immigration rates, which results in a short time lag between environmental change and the diatom response. Diatom frustules made up of silica are also very resistant to decay and thus preserve well in lake sediments.

On the basis of fossil diatom records it is possible to distinguish four general periods in the history of the ecosystem of Lake Peipsi during the last 130 years (Figure 2.14):

The pre 1950s represents a period when the ecosystem was stable and human activities in the catchment, and environmental change in general, had little influence on the ecology of the lake. Therefore this period can be considered as representing the period of natural baseline conditions. The relatively high proportion of epipsammic diatoms of the genus *Fragilaria* and epipelic *Gyrosigma attenuatum* suggests well-illuminated water column, and that light penetrated down to the lake bottom, or to the zone where periphytic diatoms might survive. The productivity of the system was low, and if compared to the values of 1980s, particularly low in plankton communities.

In the mid-1950s–1960s the state of the lake changed as demonstrated by an increase of the eutrophication indicator *Stephanodiscus parvus*, although the species

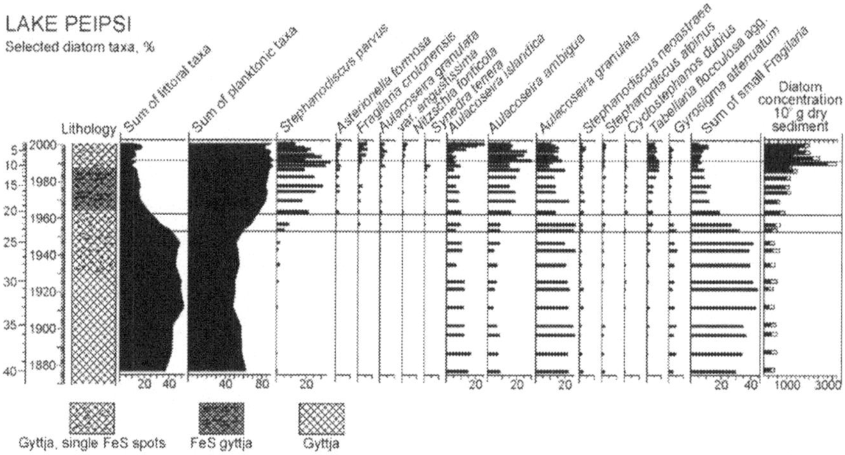

Figure 2.14 Diatom diagram of the core from Lake Peipsi s.s. In the graph to the right black bars show concentration of planktonic diatoms and white bars concentration of littoral diatoms in the sediment. The age scale left to the lithology column is based on ^{210}Pb dates and sperical fly-ash stratigraphy by Tiiu Alliksaar (from Heinsalu *et al.* submitted).

considered to indicate 'pristine association' for Lake Peipsi (*Aulacoseira granulata*, *A. islandica* and *A. ambigua*) are still dominated. The introduction of new species associated with eutrophic lakes (*Fragilaria crotonensis*, *Asterionella formosa*, *Synedra tenera* and *Aulacoseira granulata* var. *angustissima*) indicate a first step of eutrophication. The amount of littoral diatoms was reduced and the abundance of planktonic diatoms increased indicating reduced illumination of the lake bottom.

In the 1970s–1980s further changes took place in the diatom flora, indicating that the lake was eutrophied rapidly during a period of only ca. 15 years. From the 1970s the diatom assemblage of Lake Peipsi changed from conditions dominated by *Aulacoseira granulata*, *A. islandica* and *A. ambigua* to a *Stephanodiscus parvus* dominated assemblage. Also, some new eutrophic diatoms like *Diatoma tenuis* appeared.

In the 1990s sediment diatoms indicate a slight recovery of the ecosystem. Both the relative and absolute abundances of *Stephanodiscus parvus* decreased substantially during 1990s, which can reflect a reduction in the levels of phosphorus loading. The abundance of planktonic diatoms varies considerably during this period. The same pattern occurs with many planktonic diatoms, for example *Fragilaria crotonensis*, *Aulacoseira granulata* and *A.* var. *angustissima*, which implies that the ecosystem is still rather unstable.

2.5.2 Seeking reference conditions for Lake Peipsi

As there have been agricultural activities in the catchment of Lake Peipsi for centuries and over-fishing was considered a serious threat to the fish stock already by Ber (1852), it is a question, how far back in history it is necessary to go in order to find reference conditions. As increased phosphorus loading is considered the most serious and most direct anthropogenic pressure, it was considered most relevant to designate reference conditions for this element first, and to identify parameters correlating with it. To find reference conditions for TP concentration, the morpho-edaphic index (MEI), based on relationships on data from a large number of lakes in Europe and North America, was used (Vighi and Chiaudani 1985). The MEI was estimated as the ratio between alkalinity (in meq/l) and mean depth of the lake (in m). The following value of MEI for Lake Peipsi was obtained: $MEI_{alk} = 2.51/8.3 = 0.30$. For comparison, the corresponding value was calculated for another large and shallow lake located in Estonia, Lake Võrtsjärv (270 km^2, with a mean depth of 2.8 m): $MEI_{alk} = 3.04/2.8 = 1.07$. A highly significant correlation between the MEI and total P concentrations in lakes practically not affected by anthropogenic phosphorus input was discovered. This allowed the use of this relationship to calculate the reference concentration of TP for the lakes: $\text{Log}\,TP_{ref} = 1.48 + 0.33\ (\pm 0.09)\ \log MEI_{alk}$. Based on this relationship, the reference TP level for Lake Peipsi is between 18 and 23 mg/m^3 and for Lake Võrtsjärv equal to 31 mg/m^3. Comparing these values with the trophic scale of Organization for Economic Cooperation and Development (OECD) (Eutrophication …, 1982), one can see that both large lakes of Estonia fall into the mesotrophic range (Nõges and Nõges 2003). As a quality objective for Lake Peipsi the reference was set to a TP concentration multiplied by 1.5 that is close to the ecological quality ratio (EQR) suggested by Premazzi *et al.* (2003) for mesotrophic Italian lakes to meet the ecological quality objectives. Based on these considerations, the actual TP values characterising 'good'–'moderate' boundary for Lake Peipsi is equal to $TP_{good} = 1.5 \times 23 = 34$ mg/m^3. As the present average TP concentration in Lake Peipsi (44 mg/m^3) exceeds the set goal (34 mg/m^3), the water quality can be considered 'moderate' (Nõges and Nõges 2003). Considering the actual TP values measured in Lake Peipsi since 1982 (from 8 to 172 mg/m^3), they range clearly from 'high' to 'bad'.

2.5.3 Is there 'good' fish in Lake Peipsi?

In the food chain of a lake large carnivorous fishes like pike and pike-perch are the top predators placed on the highest level of the food pyramid. The amount of fish and the distribution between planktovores, benthivores and piscivores is determined by the functioning of the whole food chain. The potential productivity at all trophic levels is controlled bottom up by nutrient/food supply. The 'health' of a lake ecosystem is, however, rather determined by the efficiency of the transformation of the organic matter produced by algae through the food chain and up towards the top predators. The more efficient this transformation, the higher is the production of 'good' piscivorous fish and the better water quality can be expected. In this sense the composition of the fish community in Lake Peipsi can be considered 'good', and even improving after the increase of pike-perch since the beginning of the 1990s. The share of piscivores in the fish biomass and production has increased more than three times: from, 18% and 4.5%, respectively, in 1986 to 48% and 18%, respectively, by the turn of century (Table 2.4).

Table 2.4 Distribution of fish biomass (B) and production (P) between piscivores and non-piscivores in Lake Peipsi in 1986 and 1997–2002, calculated by A. Järvalt from the data of Kangur *et al.* (2003).

	Year – 1986		Year – 1997–2002	
	B	P	B	P
kg/ha	233	247	110	99
Piscivores (%)	18	4.5	48	18
Non-piscivores (%)	82	95.5	52	82

2.5.4 How does the food web of Lake Peipsi function?

Pelagic aquatic food chains usually consist of four trophic levels: 'phytoplankton → zooplankton → planktivorous fish → piscivorous fish'. The ecological efficiency of energy flow from one trophic level to the next is 5–20%, with a widely accepted average of 10%. Besides these pelagic food chains macrovegetation can be an important primary producer. A considerable amount of energy can also be channelled through benthic food webs in many lakes. In such systems zoobenthos plays the role of herbivores and serves as food source for bottom-feeding fish like bream, roach, etc. In a large lake like Lake Peipsi, macrovegetation does not play an important role in primary production because strong wave action limits its colonisation. Benthic fauna, first of all the invaders, *Dreissena polymorpha*, is well developed and benthic food webs are important in Lake Peipsi.

Considering the high annual production ratio (about 10%) of herbivorous zooplankton and phytoplankton, a direct relationship between these plankton groups is seen to be prevalent in the food chain of Lake Peipsi (Nõges *et al.* 1993; 2001). This means that zooplankton feeds mostly on living algae. High grazing pressure on small algae leads to the dominance of large forms in phytoplankton. The role of the 'detritus food chain' and 'microbial loop' involving bacterial consumption of dissolved organic matter and detritus, supplementing an extra link into the food chain and decreasing its efficiency seems to be rather modest in Lake Peipsi. This is confirmed also by the very small biomass of protozoan ciliates, which participate in the 'microbial loop' compared with the biomass of metazooplankton acting mainly in the 'grazing food chain' (Figure 2.15).

The prevalence of the 'grazing food chain' and the minor role of the 'microbial loop' in Lake Peipsi forms the basis of the high fish production of this lake. Phytoplankton biomass is largely 'top-down' controlled by zooplankton and also, assumingly, by bottom-inhabiting mollusc *Dreissena polymorpha*. The higher is the ratio of piscivorous to planktivorous fish, the less phytoplankton and the higher water quality are assumed.

2.5.5 How summer water blooms kill fish in Lake Peipsi

Vast amounts of excess phytoplankton biomass that the food chain is incapable to use, is built up during water blooms. Intensive photosynthesis produces large

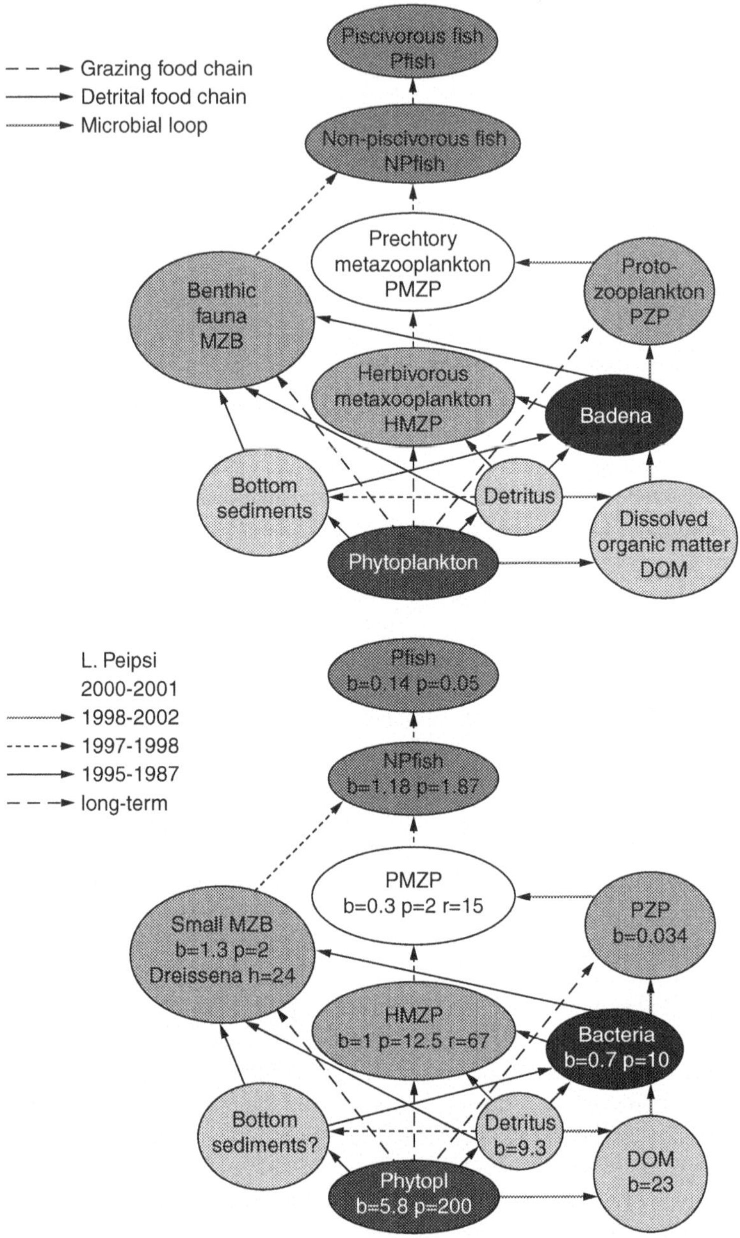

Figure 2.15 Principal and quantified structure of food web of Lake Peipsi: b = bulk, gC/m^2; p = production, gC/m^2/year; r = requested food ration, gC/m^2/year. Fish biomass and production calculated by A. Järvalt from the data of Kangur *et al.* (2003). Biomass and production of zoobentos from Timm *et al.* (2001), of zooplankton from Haberman (2001), other data from Nõges *et al.* (2001).

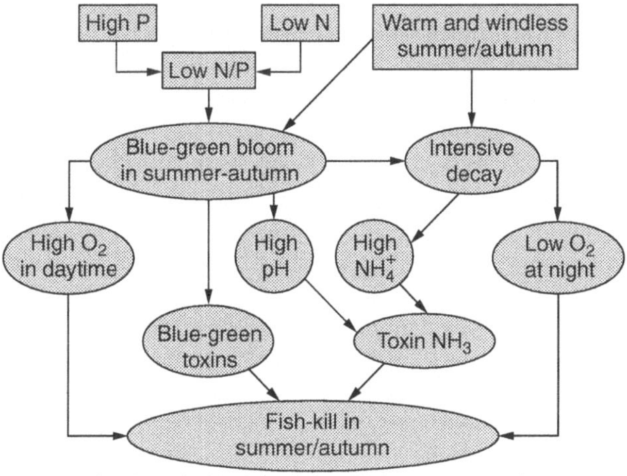

Figure 2.16 The mechanisms behind the fish-kills in summer and autumn periods in Lake Peipsi (from Nõges and Nõges 2004).

amounts of oxygen in the daytime, and if the water gets over saturated, part of the O_2 leaves to the atmosphere. At night when algal masses only consume oxygen, a deficiency may occur. Such large-scale diurnal fluctuations of oxygen concentration harms fish and makes them more susceptible to the other stress factors. High water temperature associated with water blooms makes the situation even more dangerous for fish. Other stressors accompanying water blooms are the high pH of water caused by intensive photosynthesis and elevated concentrations of ammonium released in the course of organic matter decomposition. At a pH > 9 most of ammonium is converted to toxic ammonia (NH_3) that can kill fish. Moreover, the influence of cyanobacterial toxins cannot be neglected (Figure 2.16).

Figure 2.17 In summer 2002, high water temperatures (26°C) in combination with oxygen deficiency and massive blue-green algae blooms was observed in Lake Peipsi. As a consequence, massive fish-kills were noted. For example, the maximum number of dead ruffe was 460 specimens per 1 m of shoreline at Nina village. Fish kills in Lake Peipsi have also been recorded earlier, e.g., in 1959, 1972, 1984, 1992, 1996. (Photo: Peeter Unt (Peipsi CTC).)

Water blooms and fish deaths are however not a new phenomenon in Lake Peipsi. A summer water bloom in Lake Peipsi caused by blue-green *Gloeotrichia echinulata* was reported already in 1895 by Spindler and Zengebusch (Kullus 1964). Mass development of *Anabaena flos-aquae* and *Microcystis aeruginosa* was observed already in August 1912 (Samsonov 1914). *Anabaena flos-aquae, A. spiroides* and *G. echinulata* caused water blooms in Lake Pihkva in August 1929, while in August 1934, a bloom of *Aphanizomenon flos-aquae* occurred in all three parts of the lake (Voronikhin 1950). Yearly water blooms from July to September with a maximum in August were a common phenomenon in the lake already in the 1930s (Vinkel-Voore 1935). A fish-kill during the bloom was first described in Lake Pihkva in summer 1959 (Semenova 1960), another massive smelt kill covered the whole Lake Peipsi in 1972 (Kuderskij and Fedorova 1977). Most recent fish-kills in the northern part of the lake occurred in summer 1988, in spring 1989, and in summer 2002. Most of the fish-kills coincided with blue-green blooms in summer while in 1989 a spring fish-kill was caused by the bloom of the diatom, *Aulacoseira islandica*.

2.5.6 Why do blue-green algae 'bloom' in Lake Peipsi?

The domination of cyanobacteria in the phytoplankton of the polymictic Lake Peipsi is quite ordinary as cyanobacteria are in general efficient in shallow lakes (Sheffer 1998). Water blooms can be considered also a common phenomenon for Lake Peipsi as they occurred already in the 19th century. However, the biomass of bloom-forming cyanobacteria became rather low in the conditions of heavy nitrogen loading in the 1980s and started to increase again in the post-soviet period from the 1990s onward (Figure 2.13). The new increase coincided with the decrease in riverine loading of nitrogen, and with the reduction of N/P ratio both in inflowing rivers and in Lake Peipsi (Figures 2.11 and 2.18).

Besides the N/P ratio, the concentration of phosphorus is an important factor determining cyanobacterial development. At high P levels and low N/P ratios bloom-forming blue-greens are favoured due to their ability to fix atmospheric N_2 in the conditions of N limitation. In Lake Peipsi an N/P ratio less than 30 provides an advantage for cyanobacteria (Nõges *et al.* 2002). It is quite evident that the changed proportion of N and P in loadings to Lake Peipsi have favoured cyanobacterial blooms as the N/P ratio has fallen below 30 both in rivers (Figure 2.11) and in the lake itself (Figure 2.18).

Among the bloom-forming species *G. echinulata* development starts early in the season, followed by *Anabaena*, *Aphanizomenon* and *Microcystis* later in the autumn. *G. echinulata* is generally characterised as a species requiring lower nutrient concentrations than the other indicated species (Jacobsen 1994). It has also been documented that in conditions of phosphorus limitation *G. echinulata* is able to migrate vertically in the water column taking up phosphorus at the sediment surface and using it in the euphotic zone (Pettersson *et al.* 1993; Jacobsen 1994). Due to this adaptation, *G. echinulata* is able to develop earlier in the season at lower nutrient concentrations. The N_2 fixed from the air by this species as well as the phosphorus taken up from the sediments enrich the water column with nutrients

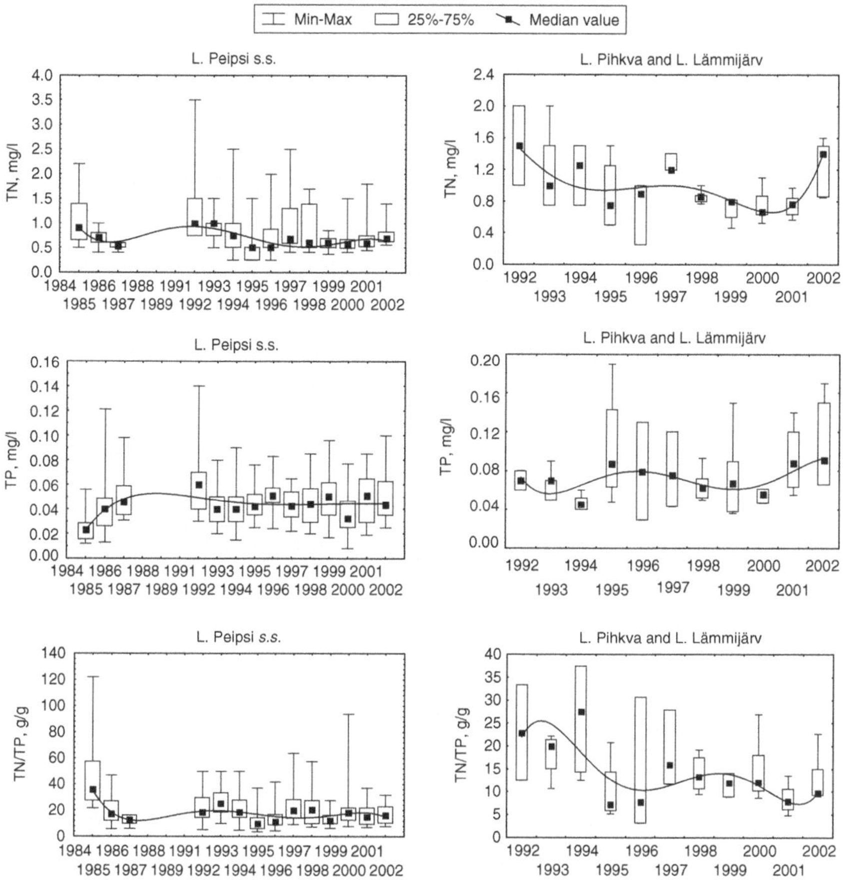

Figure 2.18 Changes of TN, TP and TN:TP ratio in Lake Peipsi 1984–2002 (from Nõges *et al.* 2005).

enabling the successive blooming of other species like *Anabaena*, *Aphanizomenon* and *Microcystis* requiring higher nutrient concentration (Nõges *et al.* 2004).

Besides nutrients, climate plays an important role in the formation of water blooms. In cold and windy summers such as in 1998, there are no blooms even at a good nutrient availability, while in warm and windless summers and autumns blooms may be extremely heavy as they were in 2002.

2.5.7 Can we prevent water blooms and fish-kills?

To reduce the risk of water blooms it is important to reduce the phosphorus load of Lake Peipsi. The reaction of in-lake P concentration to the reduced loading may be somewhat delayed due to P release from sediments. However, sediment studies have revealed that the storage of labile phosphorus compounds in Lake Peipsi sediments is much lower than in the deeper lakes and shallow lakes of smaller size (Figure 2.19). It can be assumed that in a very large lake where sediments are permanently stirred up by wind-induced waves, labile P is quickly washed out

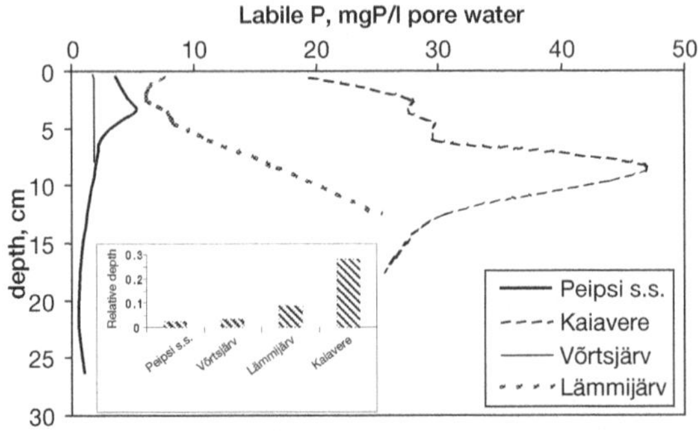

Figure 2.19 Concentration of labile phosphorus in pore water of shallow lakes of different relative depth (maximum depth/mean diameter). Unpublished data of A. Kisand are used.

from sediments into the water column and does not accumulate into the sediments so intensively as in smaller and deeper lakes. The calculation of the nutrient budget for 1998 revealed, however, that Lake Peipsi was annually retaining 50–70% of the inflowing phosphorus (Nõges *et al.* 2003b). Accumulated phosphorus seems to be firmly bound in sediments by iron and calcium compounds, and assumingly not easily released if sediments remain aerobic and pH is close to neutral. Consequently, the reduction of phosphorus loading is assumed to decrease in-lake P concentration and reduce cyanobacterial dominance and water blooms in Lake Peipsi.

In order to protect the ecosystem of Lake Peipsi and reduce the risk of water blooms and fish-kills the reduction of phosphorus loading is necessary. Protection of piscivorous fishes is suggested as a means of keeping the food chain effective and sustaining high water quality.

2.5.8 Social perceptions of the quality of the lake

A common, non-expert perception of the status of the lake would probably be that the lake is healthy if it has clear water, pleasant, open shores, and contains an abundance of 'good' (i.e. interesting for sport and commercial) fish.

Clear water indicates that there are few algae and other particles in the water, and the growth of algae is determined mainly by the available amounts of mineral nutrients – nitrogen and phosphorus. Most often phosphorus is the most important for lake quality. The amount of algae in lake water is determined both by the algae growth and by the level of consumption of algae by other organisms, as algae are eaten by small microscopic animals – zooplankton, and by some representatives of bottom fauna, for example molluscs who filter the water. If the growth of algae is balanced by consumption, then the lake water will remain clear. If growth exceeds consumption then the water will become turbid. Sometimes algae can explosively grow, forming so-called 'water bloom'. In lakes, blue-green algae (cyanobacteria) may form a water bloom that is characterised by the visible blue-green layer on the water. Such blooms make the water appear dirty and may even be toxic for swimmers and cattle. Sometimes fish-kills are accompanied by water blooms.

Lakeshores that are not overgrown by reed and other water plants are generally considered attractive. An extension of the reed belt often indicates an increased concentration of nutrients and is often most intensive close to the polluted lake inflows where nutrients enter into the lake. Although often considered a nuisance, planted areas are quite effective as water purifiers, and if it is not able to reduce the concentration of nutrients in the rivers, reed stands at the river mouths can protect the lake from eutrophication. *Aquatic macrophytes* compete with planktonic algae for nutrients and keep the water clear. However, plants can only develop in shallow areas, and in very large lakes waves restrict their growth even there.

Large carnivorous *fishes* such as pike, perch, and pike-perch that are popular with the local population and fishermen are also good for water quality. These fish eat smaller fish, for example roach, smelt, and ruffe, which normally eat zooplankton and bottom animals. Less plankivorous fish result in more zooplankton, and a more intensive consumption of algae, and therefore clearer water. Less benthivorous fish result in less disturbance of bottom sediments that they cause when looking for their prey, and therefore the water remains less turbid.

Large shallow lakes like Lake Peipsi, cannot be as clear as some other large but deeper lakes such as the Ladoga (17,872 km^2, mean depth 47 m, maximum depth 230 m), Onega (9800 km^2, max depth 110 m) and Vänern (5650 km^2 mean depth 25, max depth 106 m). The smaller the so-called 'relative depth' of the lake (the ratio of lake diameter and maximum depth), the more susceptible are the sediments to resuspension and the more naturally turbid is the water. Besides water quality and food, the spawning conditions, sheltering possibilities and the overall lake type determine the composition of fish community and the abundance of fishes.

2.6 CONCLUDING DISCUSSION

The results from the lake water quality and ecological assessment in this chapter (see Section 2.5) might also show the difficult and impossible task to find one universal criterion to diagnose all types of lakes. Each lake type has its specific features and reacts differently to environmental pressures. Instructions on how to typify lakes and how to estimate their ecological quality ('health') within different types are provided in the EU Water Framework Directive (WFD) published in December 2000 (Directive, 2000/60/EC). This is probably the most significant legislative instrument in the water field to be introduced at an international level for many years. Correct and reliable information on the status of water systems is indispensable for rational and cost-effective water management, as demonstrated by the WFD. This has general validity for all types of river basins, but is particularly relevant and problematic for transboundary water regions as clearly demonstrated in this chapter. At the same time, monitoring network data and information activities also constitute around 80% of the management activities of the various European transboundary water commissions (R. Enderlein, UN/ECE 1999, personal communication). Given the major political and economic differences (see Chapter 3) as well as variations in data availability, accessibility and quality, this has proven particularly troublesome for many transboundary water regions on the Eastern European fringe such as the Lake Peipsi drainage basin.

The results of the physical environmental assessment on Lake Peipsi given in this chapter can be summarised as:

(1) Fast eutrophication of Lake Peipsi began in the 1970s and continued up until the end of the 1980s.
(2) Since the end of the 1980s:
- nitrogen loading from Estonian catchments has decreased by about twice, reduction from Russian catchment has also been noted;
- phosphorus loadings from the Estonian side showed no decline or in some rivers a slight decrease until 1996 and after that started to increase again;
- in the largest Russian river, Velikaya, phosphorus concentration first decreased sharply at the beginning of the 1990s, but has then remained at the same (or even higher) level as in the mid-1980s;
- phosphorus concentration in Lake Peipsi s.s. has been quite stable, in the lakes Pihkva and Lämmijärv it has increased in the last few years;
- in-lake nitrogen concentration diminished up to the end of 1990s, while an increase has been noted during the last years;
- N/P ratio in loadings and in the lake has decreased to less than 30, the critical value below which the risk of blue-green blooms increases.

(3) The total amount of phytoplankton has remained at nearly the same level since 1960s.
(4) Cyanobacterial blooms and summer fish-kills have been enhanced in recent years.
(5) As regards nutrient concentration, the trophic status of Lake Peipsi has not increased in the last decade.
(6) The most acute present problems – water blooms and fish-kills are caused by changed N and P balance in loadings, that is lowering the N/P ratio into critical low levels.
(7) The destabilisation of the Lake Peipsi ecosystem is the result of decreased nitrogen loading in the conditions of non-changed, or even increased phosphorus loading.
(8) The most important measure to achieve the further improvement of water quality would be the reduction of phosphorus loading from both Estonian and Russian catchment.
(9) Protection of piscivorous fishes is important for sustaining high water quality.

REFERENCES

Ber, K.M. (1852). Issledovaniya dlya razresheniya voprosa: umen'shaetsya li kolichestvo ryby v Chudskom ozere. *Zhurnal Ministerstva gosudarstvennykh imushchestv* **43**, 248–302 (in Russian).

Directive 2000/60/EC. *Official Journal of European Commission*. L327, 1–72.

FAO (2000). *Soil and Terrain Database, Land Degradation Status and Soil Vulnerability Assessment for Central and Eastern Europe*. Land and Water Digital Media Series No. 10.

Haberman, J. (2001). Zooplankton. In: *Lake Peipsi. Flora and Fauna*, E. Pihu and J. Haberman (eds). Sulemees Publishers, Tartu, pp. 50–68.

Heinsalu, A., Alliksaar, T., Leeben, A. and Nõges, T. (2005) Sediment diatom assemblages and composition of pore-water dissolved organic matter as indicators of recent eutrophication history of Lake Peipsi. *Hydrobiologia* (submitted).

HELCOM (1994). Guidelines for the third pollution load compilation (PLC-3). *Baltic Sea Environment Proceedings 57*. Helsinki Commission.

Hirsch, R.-M. and Slack, J.-R. (1984). A nonparametric trend test for seasonal data with serial dependence. *Water Resources Research* **20**, 727–732.

Iital, A., Stålnacke, P., Deelstra, J., Loigu, E. and Pihlak, M. (2005). Effects of large-scale changes in emissions on nutrient concentrations in Estonian rivers in the Lake Peipsi drainage basin. *Journal of Hydrology* **304**, 261–273.

Jaani, A. (1996). Hydrology and water balance of Lake Peipsi. *Hydrobiologia* **338**, 11–23.

Jacobsen, B.A. (1994). Bloom formation of Gloeotrichia echinulata and Aphanizomenon flos-aquae in a shallow, eutrophic, Danish lake. *Hydrobiologia* **289**, 193–197.

Kangur, P., Kangur, A., Kangur, K. and Möls, T. (2003). Condition and growth of ruffe Gymnocephalus cernuus (L.) in two large shallow lakes with different fish fauna and food recourse. *Hydrobiologia* **506**(1), 435–441.

Kuderskij, L.A. and Fedorova, G.V. (1977). Snizhenie zapasov snetka v bol'shikh vodoemakh Severo-Zapada Evropejskoj chasti SSSR v 1973–1975 gg. *Rybokhozyajstvennoe izuchenie vnutrennikh vodoemov* **20**, 3–8 (in Russian).

Kullus, L. (1964). Peipsi-Pihkva järve uurimisest ajavahemikul 1950–1917. *Eesti Geograafia Seltsi Aastaraamat 1963*, Tartu, pp. 148–158.

Laugaste, R., Nõges, T., Nõges, P., Yastremskij, V., Milius, A. and Ott, I. (2001). Algae. In: *L. Peipsi. Flora and Fauna*, E. Pihu and J. Haberman (eds). Sulemees, Tallinn, pp. 31–49.

Libiseller, C. and Grimvall, A. (2002). Performance of partial Mann–Kendall tests for trend detection in the presence of covariates. *Environmetrics* **13**, 71–84.

Loigu, E. and Vassiljev, A. (1997). Evaluation of water quality response to sudden changes in the amount of fertilizers used in Estonia. In: *Hydrology and Environment. Proceedings of the Baltic States Hydrology Conference*, Kaunas, pp. 123–130.

Loigu, E., Leisk, Ü. and Blinova, I. (1995). The state of aquatic environment. Inland water quality. In: *Protection of the Coastal Wetland Area – Matsalu Catchment Area*. Tallinn, Stockholm, 23–32.

Löfgren, S., Gustafson, A., Steineck, S. and Stålnacke, P. (1999). Agricultural development and nutrient flows in the Baltic States and Sweden after 1988. *Ambio* **28**(4), 320–327.

Mander, Ü., Kull, A., Kuusemets, V. and Tamm, T. (2000). Nutrient runoff dynamics in a rural watershed: influence of land-use changes, climatic fluctuations and ecotechnological measures. *Ecological Engineering* **14**(4), 405–417.

Nõges, P. and Nõges, T. (2003). Monitoring of large shallow transboundary Lake Peipsi (Estonia/Russia) and the indicators and criteria worked out for assessment of its ecological status. *TemaNord*. **547**, 71–80.

Nõges, T. and Nõges, P. (2004). Consequences of catchment processes and climate changes on the ecosystems of large shallow temperate lakes. In: *Proceedings of the 7th International Symposium on Fish Physiology, Toxicology, and Water Quality*, G.L. Rupp and M.D. White (eds). Tallinn, Estonia, May 12–15, 2003, US. Environmental Protection Agency, Ecosystems Research Division, Athens, GA, USA, EPA 600/R-04/049, pp. 341–350.

Nõges, T., Haberman, J., Timm, M. and Nõges, P. (1993). The seasonal dynamics and trophic relations of the plankton components in Lake Peipsi (Peipus). *Internationale Revue Der Gesamten Hydrobiologie* **78**, 513–519.

Nõges, T., Haberman, J., Kisand, V., Laugaste, R. and Zingel, P. (2001). Trophic relations and food web structure of plankton community in Lake Peipsi s.s. In: *L. Peipsi. Flora and Fauna*, E. Pihu and J. Haberman (eds), pp. 74–81.

Nõges, T., Blinova, I., Jastremski, V., Laugaste, R., Loigu, E., Skakalski, B. and Tõnno I. (2003b). Reduced nitrogen loading enhance cyanobacterial blooms in Lake Peipsi. In: F. Bernardini, M. Landsberg-Ucziwek, S. Haunia, M. Adriaanse, R. Enderlein (eds) *Proc. of 2nd Int. Conf. Sustainable Management of Transboundary Waters in Europe*. Poland, Miedzydroje, 21–24 Apr. 2002. 501–504.

Nõges, P., Nõges, T., Haberman, J., Kangur, K., Kangur, A., Kangur, P. Laugaste, R., Mäemets, H., Ott, I., Timm, H., Yastremskij, V.V. and Virro, T. (2003a). Water quality criteria according to Water Framework Directive: how they work in the largest Transboundary lake in Europe, L. Peipsi? In: F. Bernardini, M. Landsberg-Ucziwek, S. Haunia, M. Adriaanse, R. Enderlein (eds) *Proc. of 2nd Int. Conf. Sustainable Management of Transboundary Waters in Europe*. Poland, Miedzydroje, 21–24 Apr. 2002. 393–401.

Nõges, P., Leisk, Ü., Loigu, E., Reihan, A., Skakalski, B. and Nõges, T. (2003). Nutrient budget of Lake Peipsi in 1998. *Proc. Acad. Sci. Estonia. Ser. Ecology* **4**, 407–422.

Nõges, T., Tõnno, I., Laugaste, R., Loigu, E. and Skakalski, B. (2004). The impact of changes in nutrient loading on cyanobacterial dominance in Lake Peipsi (Estonia/Russia). *Archiv Fur Hydrobiologie* **160**(2), 261–279.

Nõges, T., Laugaste, R., Loigu, E., Nedogarko, I., Skakalski, B. and Nõges, P. (2005). Is the destabilisation of Lake Peipsi ecosystem caused by increased phosphorus loading or decreased nitrogen loading? *Water Science and Technology* **51**, 3–4, 267–274.

OECD (1982). *Eutrophication of water, monitoring, assessment and control.* Organization for Economic Cooperation and Development (O.E.C.D.), Paris, p. 150.

Oláh, J. and Oláh, M. (1996). Improving landscape nitrogen metabolism in the Hungarian lowlands. *Ambio* **25**, 331–335.

Ott, I. (2001). Typology and ecological classification of Estonian Lakes. In: *Classification of Ecological Status of Lakes and Rivers*, S. Bäck and K. Karttunen (eds). TemaNord 584, Nordic Council of Ministers, Copenhagen, pp. 52–63.

Pettersson, K., Herlitz, E. and Istvánovics, V. (1993): The role of *Gloeotrichia echinulata* in the transfer of phosphorus from sediments to water in Lake Erken. *Hydrobiologia* **253**, 123–129.

Pihu, E. and Kangur, A. (2001). Fishes and fisheries management. In: *L. Peipsi. Flora and Fauna*, E. Pihu and J. Haberman (eds). Sulemees, Tallinn, pp. 100–111.

Premazzi, G., Dalmiglio, A., Cardoso, A.C. and Chiaudani, G. (2003). Lake management in Italy: the implications of the Water Framework Directive. *Lakes and Reservoirs: Research and Management* **8**, 41–59.

Samsonov, N.A. (1914). Plankton Pskovskogo vodoema. II. Vesennij i letnij plankton. *Trudy Promyslovo-Nauchnoj Ekspeditsii Po Izucheniyu Pskovskogo Vodoema* **1**(4), 1–18 (in Russian).

Scheffer, M. (1998). *Ecology of Shallow Lakes*. Chapman and Hall, London.

Semenova, N.I. (1960). *Prichiny gibeli snetka v Pskovskom ozere v avguste 1959 goda. Naychno-tekhnicheskij byulleten' GosNIORKh* **10**, 23–24 (in Russian).

Starast, H., Milius, A., Möls, T. and Lindpere, A. (2001). Hydrochemistry. In: *Lake Peipsi. Meteorology, Hydrology, Hydrochemistry*, T. Nõges (ed.), Sulemees, Tallinn, pp. 97–131.

Stoermer, E.F. and Smol, J.P. (eds) (1999). *The Diatoms: Application for the Environmental and Earth Sciences*. University Press, Cambridge, p. 469.

Stålnacke, P. and Grimvall, A. (2000). Hydrological normalization of nutrient deliveries from agricultural catchments. Reviewed Proceedings from *Eleventh Annual Conference on Applied Statistics in Agriculture*, Kansas State University, pp. 145–155.

Stålnacke, P., Sults, Ü., Vasiliev, A., Skakalsky, B., Botina, A., Roll, G., Pachel, K. and Maltsman, T. (2002). Assessment of riverine loads of nutrients to Lake Peipsi, 1995–1998. *Archiv Fur Hydrobiologie* **3–4**(Suppl. 141), 437–457.

Stålnacke, P., Loigu, E., Melnik, M., Nõges, T. and Vetemaa, M. (2005). *Transboundary diagnostic analysis of lake Peipsi/Chudskoe*. NIVA Report No. 4956-2005, p. 62.

Timm, T., Kangur, K., Timm, H. and Timm, V. (2001). Zoobenthos. In: *L. Peipsi. Flora and Fauna*, E. Pihu and J. Haberman (eds), Sulemees, Tallinn, pp. 82–99.

Tonderski, A. (1997). *Control of nutrient fluxes in large river basins*. Ph.D. Thesis, Linköpings Studies in Art and Science 157, Sweden.

Unwin, T. (1997). Agricultural restructuring and integrated rural development in Estonia. *Journal of Rural Studies* **13**(1), 93–112.

Vagstad, N., Stålnacke, P., Andersen, H.E., Deelstra, J., Jansons, V., Kyllmar, K., Loigu, E., Rekolainen, S. and Tumas, R. (2004). Regional variations in diffuse nitrogen losses from agriculture in the Nordic and Baltic region. *Hydrology and Earth System Sciences* **8**(4), 651–662.

Vassiljev, A. (1994). Modeling wash-off and leaching of pollutants by spring-time flow. *Journal of Hydrology* **159**, 215–222.

Vighi, M. and Chiaudani, G. (1985). A simple method to estimate lake phosphorus concentrations resulting from natural background loadings. *Water Research* **19**, 987–991.

Vinkel-Voore, R. (1935). Vee õitsemine Peipsi järvel. *Eesti Loodus* **1**, 24.

Voronikhin, N.N. (1950). Fitoplankton Pskovskogo vodoema. In: *Sporovye rasteniya*, Vol 5. Moskva, Leningrad, pp. 5–19 (in Russian).

3

The administrative and political environment

J. Johansson and L. Kalashnyck

3.1 INTRODUCTION

Chapter 2 described the physical attributes of the Lake Peipsi region; in this chapter political, socio-economic, and policy-relevant aspects of managing the environment are presented. The descriptions and analysis of the administrative and political aspects of the region are based on the results of surveys, focus groups, interviews with members of the public and stakeholders, content analyses of documents, studies of mass media sources, and interviews with environmental authorities on both the Estonian and Russian sides of the Estonian–Russian border. Especially important sources of information were the semi-structured interviews with officials working with environmental issues. These interviews concentrated on institutional aspects of water management and included questions aimed at elucidating views on the most important forms of environmental problems, the concepts behind environmental management and policy, the organisation of regional environmental decision-making, the barriers to effective environmental management, how and by whom environmental management should be managed, relations between different environmental

 Integrated Transboundary Water Management in Theory and Practice: Experiences from the New EU Eastern Borders edited by Geoffrey D. Gooch and Per Stålnacke. ISBN: 1843390841. Published by IWA Publishing, London, UK.

authorities, and perceptions of the resilience of the environment. The interviews also examined the ways in which transboundary environmental management was managed, and the factors that hindered or facilitated environmental co-operation between the two countries. A final section aimed at formulating future scenarios. The interviews and other forms of data collection were recorded, translated, and transcribed.

3.2 THE BALTIC STATES

The region that our study focuses on is to the east of the Baltic Sea. The three Baltic States are, from north to south, Estonia, Latvia, and Lithuania. All three were independent in the 1920s and 1930s, but were incorporated into the Soviet Union as Soviet Republics in the 1940s. While they are often referred to as if they formed a unified unit, there are in fact substantial differences between them. The Estonian language belongs to the Finno–Ugric group of languages, and has such considerable similarities with the Finnish language. This was advantageous for Estonia both before and after the country's emergence from the Soviet Union, and Estonia's economic development during the early 1990s was helped by near contacts with Finnish experts and businessmen. Latvian and Lithuanian belong to other language groups, and historically, Lithuania was part of the Polish–Lithuanian power constellation. During the Soviet period, Russian was a common language used for communication between these three countries, but after independence English was more likely to be used for this purpose. The large percentage of ethnic Russians during the Soviet period, especially in Latvia (ca. 40%) and Estonia (ca. 25%) was the result of population and industrialisation strategies implemented by the Soviet authorities. During this time, ethnic conflicts were not officially acknowledged (Glenny 1993) although they did exist. The Baltic struggle for independence at the end of the 1980s and beginning on the 1990s was strongly influenced by ethnic and national sentiments, and the position of the ethnic Russian minorities radically changed. Estonian, Latvian, and Lithuanian became the official languages of the Baltic States and many ethnic Russians in these countries were unable to speak them.

The fate of ethnic Russian living outside of the Russian Federation became an important issue for the new Russian state, as about 17% of the total Russian population of the former USSR suddenly found themselves living in the 'near abroad' (Sakwa 2002). Almost 800,000 of these lived in Latvia and almost 500,000 in Estonia at that time. Ethnic Russians were often equated with the Soviet system (Bremmer and Taras 1997) and disputes over citizen rights during the early 1990s amplified the feelings of many Russians, both in the Baltic States and in the Russian Federation, that ethnic Russians were discriminated against. The lack of political influence of ethnic Russians in the new states also exuberated these sentiments (Melvin 1995). The situation was not however completely conflictual. Many ethnic Russian did not oppose the move to independence, and most conflicts have been successfully managed during the 1990s and early 2000s. Environmental and water management in the Lake Peipsi drainage basin should however be seen with the political and administrative history of the region in mind.

3.3 ENVIRONMENTAL MANAGEMENT IN RUSSIA DURING THE EARLY 2000S

Although the Soviet Union was officially a federal system, it was in fact a centralised state in which the government in Moscow determined to a large extent policy in the provinces (Kelley 1999). In the Soviet system, the Union Republics were divided into *oblasts* (provinces or regions), and *raions* (districts). The system was hierarchical; towns and villages were subordinate to *raions* (districts), *raions* and cities were subordinate to the *oblasts* (regions), and the *oblasts* (regions) subordinate to the Union Republics. The current territorial organisation of the Russian Federation reflects to a certain extent this old Soviet system (Kourlindskaia and Kachanova 2002).

The provinces aspired for, and gained, more autonomy during Gorbachev's *perestroika*, and Yeltsin encouraged the regions to strive for privileges in the areas of, for example, tax collection and natural resource management (Evangelista 2000). During Yeltsin's presidency, some regions developed authoritarian regimes (Klyamkin and Shevtsova 1999) but the President of the Russian Federation, Vladimir Putin's policy has, after 2000, involved a re-centralisation of the political and administrative system. In May 2000, President Putin issued a decree that re-organised the country into seven federal districts, each run by a presidential appointee. Russia then had three levels of government: federal, regional, and local. There were a total of 89 'subjects of federation', the most common of which was the *oblast*, or region. Most of these regions had an elected governor and their own administration and legislature. There were however appreciable differences between the powers and composition of the regions (Sakwa 2002, p. 224). This re-organisation may have resulted in increasing the size of the central bureaucracy without improving efficiency (Stoner-Weiss 2002), and the central government seemed to have had problems in implementing its policies in the regions (Gelman 2001). The system also involved the division of regions into *donors* or *recipients* of state financial subsidies. Only a few of the regions were donors, most were recipients (Sakwa 2002, pp. 237–238) and agricultural regions such as those on the north-west borders were dependent on state subsidies (Shlossberg 2002). The new Tax Code of 2001 also shifted the financial balance in favour of the federal centre (Petrov 2001; Kuzmin and Kachanova 2002). The local authorities were however expected to finance and deliver a broad range of public services (Brown 2001), and they were fully responsible for water and sewage services, as well as waste disposal and sanitation, although they often encountered difficulties in providing services due to limited resources (Sakwa 2002, p. 251). These difficulties originated in a lack of municipal financial independence resulting from a system in which local budgets were fully dependent on regional governments, which in turn received transfers from the federal centre (Kuzmin and Kachanova 2002, p. 256; Sakwa 2002, p. 250). Functions were often delegated to the local authorities without the necessary financial support, and regional governments sometimes deprived the municipalities of funding (Kourlindskaia and Kachanova 2002, p. 197). The amount received by the municipalities can therefore depend on the negotiating abilities of municipal managers (Kuzmin and Kachanova 2002, p. 529).

Local interest groups and political parties often played only a minor role in the policy-process (Kourlindskaia and Kachanova 2002, p. 180), and although the Law on the General Principals of Local Self-Government gave citizens the right to hold local referenda and to initiate legislation on local issues, these rights were seldom used (Kourlindskaia and Kachanova 2002, p. 182). The ability of local authorities to design their own policies was further complicated by the fact that both the President of the Russian Federation and the governors of the regions had the right to dismiss local governments if their decisions violated federal or regional legislation (Kuzmin and Kachanova 2002, p. 529).

The causes of environmental degradation in the Soviet Union have been characterised as the results of the dominant ideology, rapid, concentrated and heavy industrialisation processes, the militarisation of the Soviet economy, the negative effects of the command economy, and weak environmental protection safeguards (Shaw 1999; Ivanova 2001; Peterson and Bielke 2001; Wernstedt 2002). Others have pointed out the absence of politics as an aggravating factor (Barry and Frankland 2002, p. 406), that Soviet ideology was technological rather than ecological, and that the environment was seen as a resource to be exploited (Shaw 1999; Sakwa 2002). The elaborate system of very strict environmental regulations that was developed during the Soviet period was also in practice not effectively enforced (Shaw 1999, p. 133; Ivanova 2001, p. 7). The break-up of the Soviet Union resulted in changes in environmental management (although perhaps not in the beliefs of the administrators), and the Yeltsin administration initiated an advanced environmental legislative framework (Shaw 1999, pp. 143–149; Sakwa and Stevens 2000; Peterson and Bielke 2001, p. 67; Sakwa 2002, p. 329). However, the Ministry of Environmental Protection and Natural Resources created by the Yeltsin administration in 1991 was transformed into the State Committee for Environmental Protection in 1996, with a lower status than that of a ministry (Peterson and Bielke 2001, p. 67). In May 2000, President Putin disbanded the agency and transferred its staff and functions to the Ministry of Natural Resources of the Russian Federation (Massa and Tynkkynen 2001, p. 15). While the State Committee for Environmental Protection may during its time mainly have 'served as a symbol of Russia's desire to appear like a developed and modern country' (Peterson and Bielke 2001, p. 68), it remains to be seen how effective environmental interests will be managed in the Ministry of Natural Resources. The environmental management system in Russia at the time could be characterised as a mixture composed of almost all existing branches of power. Both the lower chamber of the Russian Parliament (the *State Duma*) and the upper chamber (the *Federation Council*) had committees concerned with environmental issues. Water issues at the federal level were the responsibility of the State Water Service, which was divided into three Directorates; the Directorate of Water Management, the Directorate of Basin Planning, and the Directorate of Water Resources (Russian Federation Ministry of Natural Resources 2002). These were responsible for water provision and the water budget, transboundary waters and the water fund, water management programmes and policies, basin agreements, regulation and exploitation of water management systems, flood prevention and the safety of hydro-electric facilities, the operative planning of water management activities, the state water cadastre, stock-taking of water uses, and water quality recuperation (Budarin *et al.* 2001, p. 18).

The result of the complicated and inter-dependent nature of Russian environmental management was that 'environmental policy-making appears fragmented and many of the institutions responsible for it are unstable' (Wernstedt 2002, p. 1), a view shared by other critics of the system (Yablokov 2001, p. 3). Russian environmental management was also handicapped by the regulatory system based on an extensive collection of subordinate legislative documents (decrees, resolutions, regulations, administrative orders, decisions, etc.). This Byzantine system demanded time, resources, and knowledge in order to be understood, which many communities lacked (Ivanova 2001, p. 11). The unrealistic stringency of the old environmental standards system mentioned above also led to non-compliance by industrial and business interests (Ivanova 2001, p. 14) and corruption among state and local officials (Massa and Tynkkynen 2001, p. 14). As regional and local authorities were primarily responsible for the enforcement of environmental protection, and it was precisely these administrative levels that lacked sufficient funds, these authorities were also susceptible to the influence of strong and well-organised interest groups (Ivanova 2001, p. 13). By providing services, the business community also attempted to secure preferential treatment (Wernstedt 2002, p. 14).

The overall result was a system for environmental management characterised by a complicated institutional framework, agencies with overlapping functions, poor co-ordination, and interagency conflict (Ivanova 2001 p. 11). As a result, local administrations sidelined decisions taken by the federal and regional environmental authorities on the grounds of economic necessity (Crotty 2002). The Committee of Natural Resources was the territorial arm of the central governmental Ministry of Natural Resources in the Pskov region. The Committee was subdivided into sections concerned with environmental protection, water management, forestry, and mineral resources. The Federal Water Management Administration was also co-ordinated by the Committee, and was responsible for setting and licensing water use quotas, monitoring water objects, the oversight of hydro-technical facilities, and the analyses of discharges into water bodies. Another important federal agency was the Federal Land Cadastre Service of Russia, which regulated and oversaw land legislation and use, and was organised in 24 districts and 2 city branches. The use of fish resources was controlled and administered by the Regional Fish Inspectorate, which was under the State Committee of Fisheries in Moscow. The Regional Fish Inspectorate was organised in three branches, and monitored fish resources. It was given the task of monitoring water pollution and quality, but that assignment was not at the time fully developed. Another institution involved in the management of fish resources was the regional branch of the State Research Institute of Lake and River Fishery. As a result of the post-Soviet institutional reforms it was answerable to the Ministry of Agriculture. Finally, the Hydrometeorology and Environmental Monitoring Centre of the Pskov Region represented the Federal Service for Hydrometeorology and Environmental Monitoring at the regional level. The Environmental Prosecutor's Office of the Pskov Region had judicial responsibility for environmental actions, and was subordinate to the Prosecutor

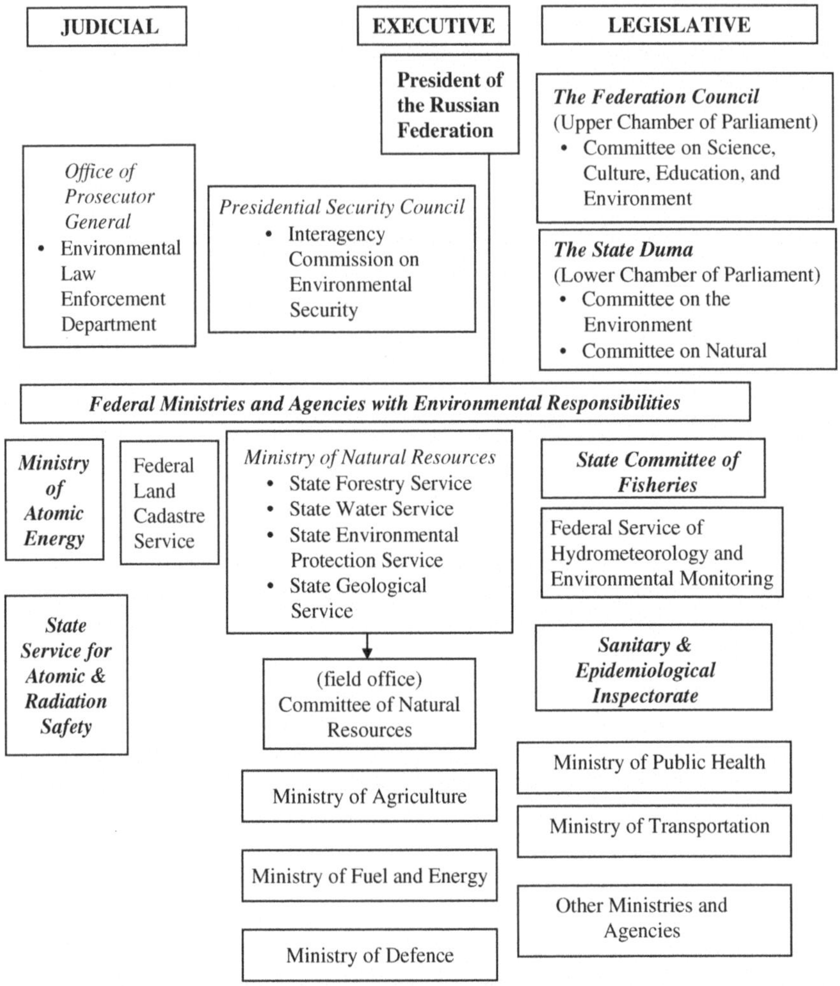

Figure 3.1 The structure of environmental management in the Russian Federation (federal level) in 2003. *Source:* Kalashnyck (2003).

General of the Russian Federation. Much of the environmental management at the municipal level was organised together with the other authorities (Turnock 2001, p. 176). The municipal level was, however, according to Russian law, responsible for domestic waste disposal, the supply of drinking water and wastewater treatment, sanitary conditions, the oversight of land use, water regulation and the exploitation of non-strategic resources. According to the officials interviewed the most acute problems in the *oblast* were a result of insufficient wastewater treatment.

3.3.1 The Pskov region and environmental management

The Pskov region is a region in the north-west of Russia bordering on Estonia, Latvia, and Belarus. The region is sub-divided into 24 districts with 9 of them lying on the region's fringes. The regional capital and largest city is the city of Pskov. Although within Russia the Pskov region borders on the two highly developed regions, Novgorod and Leningrad *oblasts*, the region continues to be an agricultural periphery. During the early 2000s the region was characterised as economically depressed with high unemployment rates, outward migration of human resources, and deteriorating demographic trends. From the environmental perspective, the region was not richly endowed with natural resources. However, there were significant fish and forest resources, which were quite actively exploited. Lake Peipsi on the border with Estonia was, and still is, considered one of the best in Europe in terms of commercial fishing and the wetland zones around Lake Peipsi are under the protection of the international Ramsar Convention as an important biodiversity resource.

3.3.2 The administrative setting

The bureaucratic complexity in managing environmental policy in the Pskov region was a result of the formal division of powers among the environmental agencies at the federal level. The Committee of Natural Resources, field division of the central governmental Ministry of Natural Resources in the Pskov region, played a central role in the co-ordination and oversight of the implementation of environmental policy in the region. Like the Ministry, the Committee was sub-divided into Environmental Protection, Water Management, Forestry and Mineral Resources Sections. In addition, the Committee's Department of State Environmental Oversight oversaw the implementation of environmental decisions and projects, and fed information back to the Committee's senior management. This Committee also co-ordinated the Federal Water Management Administration charged with setting and licensing water use quotas, monitoring water objects, the oversight of hydro-technical facilities, and the analyses of discharges into water bodies. It should be emphasised that although the Committee operated at the regional level, it represented the interests of the central government.

Another important environmental actor was the Committee of Land Resources subordinated to the Federal Land Cadastre Service of Russia. The Committee's key function was to regulate land cadastre, that is, registration. In addition, the Committee of Land Resources oversaw the enforcement of land legislation and land use. Considering the fact that the number of land users in the region was estimated at some 380,000, the Committee's role in environmental affairs was hard to overrate. This agency had 24 district and 2 city branches affiliated with municipal governments so as to jointly manage land issues in municipalities.

The exploitation of fish resources was controlled and overseen by the Regional Fishing Inspectorate subordinated to the central governmental State Committee of Fisheries. The Regional Fishing Inspectorate was sub-divided into three branches to monitor fish resources and to enforce fisheries regulation. It was assigned to monitor water pollution in the lakes where fisheries operate, yet this capacity remained to be fully developed. Placed under another ministerial roof, that of the Ministry of Agriculture was the State Research Institute of Lake and River Fishery. Its key function was to monitor and research fish resources in the region's lakes in order to prevent their over-exploitation. In view of the Pskov region's position on the shore of one of Europe's largest lakes, the role of both institutions in terms of the formulation of regional policy on fisheries was very important.

A number of other federal environmental agencies such as the Federal Service for Hydrometeorology and Environmental Monitoring, the Sanitary and Epidemiological Inspectorate, the Ministry of Public Health, etc. also had their field offices in the Pskov region. The judicial oversight of environmental actions was the responsibility of the Environmental Prosecutor's Office of the Pskov region, which was subordinated to the Prosecutor-General of the Russian Federation. Apart from the central governmental environmental agencies operating in the region, the other two key players pointed out by the environmental managers were the Regional Government and the Local (Municipal) Governments. In fact, there were 26 municipalities in the Pskov region, but they are regarded here as one actor in relation to other environmental actors.

According to Russian law of environmental protection and the administration of natural resources is the joint remit of the central or *federal* government and the regional government. The executive branch of the regional government, the *Oblast Administration (OA)*, was headed by the Governor and had several departments charged with environmental functions. The Department of Mineral Resources and License Policy was tasked with issuing licenses and permits for natural resources' exploitation on the territory of the region. This was administered together with the Committee of Natural Resources, which represented the central government. The Department of Nature Use managed a broad spectrum of environmental and economic issues including fisheries and timber companies. The executive branch of the Regional Government was organised into different committees and departments responsible for health, housing, energy and agricultural policies, state property, etc., many of which also had an environmental dimension. The legislative branch of the regional government, the Legislative Assembly (LA) of the Pskov region, or the *Regional Duma*, was characterised by several of the managers as having little involvement in tackling environmental issues. Reportedly, the LA did not even have an environmental committee at that time. Some managers noted that the executive branch in the Pskov region had been able to keep the influence of the legislative branch in check, at least in certain areas. Reportedly, after the victory of the governor's party in the elections to the Regional Assembly during that period, the Assembly fell under control of the governor.

Municipal governments in Russia were charged with several important environmental functions such as waste disposal, water- and sewage-water treatment, land and environmental protection. Many of the environmental functions were administered in co-operation with the field offices of the central governmental environmental agencies operating in the region. According to the environmental managers, the main environmental problem in the Pskov region at the time was inadequate wastewater treatment. Wastewater treatment facilities both in the regional centre Pskov and smaller border towns of Pechory and Gdov were said to be badly in need of serious upgrading, both in the case of municipal wastewater discharges and industrial discharges. According to one manager, the urgency of the problem in

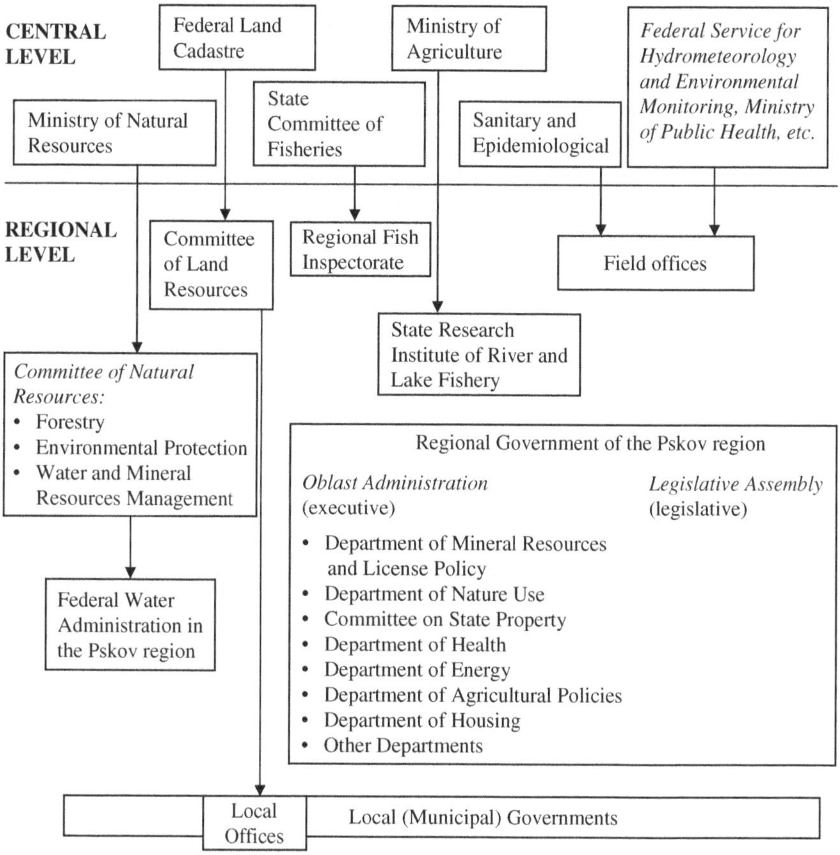

Figure 3.2 Structure of environmental management at the regional level in 2003 (arrows stand for lines of subordination). *Source:* Kalashnyck (2003).

one of the municipalities was extremely high because municipal wastes were reportedly dumped directly into water bodies without any treatment. It was also not uncommon that water supply systems left from the Soviet times were in need of serious repair. In combination, these factors caused deterioration of drinking water quality in the regional centre with a population over 200,000 inhabitants. Another serious environmental ramification of this problem, according to one of the managers, was the continuing eutrophication of Lake Peipsi, as a result of 'the increase in algal production in the lake and other changes caused by increased nutrient loadings from wastewater and agriculture'. The quality of water resources in general and drinking water in particular was said to be one of the gravest concerns not only in the region, but in the whole of Russia.

Waste disposal and processing was perceived by a considerable number of managers to be another serious environmental problem. In the city of Pskov, this problem had several dimensions. Firstly, the city waste management was running into problems as the large city depended on the outdated storage facilities that could not effectively cope with the increasing waste generation. Secondly, the fact that domestic wastes entered the waste disposal system without being sorted out greatly complicated its processing. Thirdly, recycling of reusable types of domestic waste such as paper remained unsatisfactory. Quite a similar picture concerning inadequate waste disposal, according to three managers, existed in the other municipalities. For example, in the Gdov municipality with its long shoreline spontaneous dumps resulting from unorganised tourism presented an acute problem. While Russia adopted a modern waste management policy approach in the late 1990s, as embodied in the Federal Waste Programme, improvement of waste management in the region was still unsatisfactory, mainly because of financial constraints.

The third major environmental problem in the early 2000s had to do with the fishing in Lakes Peipsi and Pskovskoye. The managers shed light on several aspects of this issue, which are addressed in more detail later. In brief, according to two managers, the number of Russian fishing enterprises continued to grow, as the market mechanisms to make economically unviable fisheries leave the business were not properly in place. As a result, the pressure on fish resources and the lakes' fauna/flora in general continued to increase. Organised illegal fishing continued to proliferate, as well as the tendency of even legitimate fishing enterprises to under-report the real amounts of the caught fish to evade taxation burdens. These led to further problems with discrepancies in official statistics. In addition, the occasional use of trawl-nets reported by one of the managers may have seriously damaged the underwater flora of Lake Peipsi. As the regional government reportedly turned to doing something about the unchecked exploitation of fish resources only in 2001, this constituted one of the gravest potential environmental problems.

Though air pollution was not mentioned as a primary problem, some managers noted its harmful potential in urban areas. This perspective was confirmed by the statements of the State Sanitary and Epidemiological Inspectorate charged with monitoring air and water quality. The Inspectorate registered a tendency for air pollutants to increase in the industrial zones in the city of Pskov. It was also pointed that the technical capabilities of the labs in the regional centre were far from comprehensive, and unable to register some dangerous types of air pollutants.

Evidently there are other *point* sources of pollution in the Pskov region. For example, the leaking underground petroleum reservoirs that belonged to the Soviet Air Force caused pollution of underground water in the proximity of the regional capital. It was a subject of continuing negotiations between the municipal authorities and the military, and a focus at the time of environmental NGO's activities to improve the situation. However, as the objective of this chapter is to present a general picture of environmental management problems in the region, a more detailed discussion of these problems falls outside the scope of the study.

The major regional city of Pskov, as well as the smaller towns of Pechory and Gdov, were in need of significant improvements in wastewater treatment. Waste disposal was also seen as a major problem, partly because of outdated waste storage facilities, partly because of a lack of waste segregation, and partly because of waste left in the countryside by unorganised tourism. However, one of the most important sources of pollution, diffuse pollution from agriculture, was hardly mentioned by the managers. This may be due to lack of awareness, but also to the decreased volume of agricultural production in the region. Some of the Russian managers believed that the decrease in agricultural production that had occurred on both sides of the border since the beginning of the 1990s had reduced environmental problems such as diffuse agricultural pollution. Concerning environmental management, the main concepts behind environmental policy and management formulated at the federal level were not well understood by the Russian managers in the *oblast*. Many could not, or would not, discuss this issue. The Russian environmental managers were also unsure about how environmental decisions were taken at the federal level. A general opinion was that the implementation process was hierarchical, based on strong vertical subordination, and that the main principal was '*to be under the single command*' of the Minister (Russian manager). Procedures at the regional level were seen as '*excessively regulated*', and at the municipal level it was felt that '*the path from an idea to its implementation involves too many people*' (Russian manager). At the regional level, decision-making appears to be '*quite closed*', and '*whether any environmental decisions emerge at all is dependent on . . . the economic interest in this*' (Russian manager). At the local level, a common perception was that it was better to do nothing at all than to risk making a wrong decision. The problems of synchronisation was also noted by the Russian environmental managers, and it was pointed out that the '*former Soviet lack of co-ordination between the agencies has persisted and has yet to be resolved*' (Russian manager). The opinions of the interviewees confirmed that the major problem with environmental management was lack of funding. Whereas 60% of the funding generated by pollution payments used to remain in the municipalities, it was felt that the regional and federal authorities now appropriated this funding at this time. The requisition of federal funds by the municipalities was a process characterised by delays and reductions. One interviewee claimed that '*they* (regional governments) *just don't give a damn about the raion*', and another stated that the funding situation made it necessary to '*snatch money for solving environmental problems*' (Russian manager). The rhetorical nature of the environmental debate was described as '*we just keep on talking about the environment . . . but there are few deeds*' (Russian manager), and the lack of public interest was seen as a major problem. One interviewee claimed that '*until the citizen realises that garbage should be put into a garbage can, not dumped in a field or in the woods, all conversations about enlightenment and environmental protection will remain only conversations*' (Russian manager).

3.4 ENVIRONMENTAL MANAGEMENT IN ESTONIA IN THE EARLY 2000s

Estonia regained its independence on 20 August 1991 following the break-up of the Soviet Union. On the 28 June 1992, the Fourth Constitution of the Republic of Estonia was adopted. The head of the state is the president who is elected by the parliament or by an electoral body if necessary. The electoral body is made up of members of the Riigikogu and representatives of the local governments. The

Figure 3.3 Lake Peipsi is one of the best fishery water bodies in Europe. High fish catches (about 30–40 kg/ha a year) and high qualitative composition of ichthyofauna, e.g., over 50% of total annual catch is represented by species like smelt, pikeperch, bream, perch, pike, burbot, whitefish and earlier also vendace. Ice-fishing in Lake Peipsi is very popular on the lake Peipsi. (Photo: Peeter Unt (Peipsi CTC).)

legislative body of the Republic of Estonia is the parliament, *Riigikogu*. The supreme executive power is vested in the Government of the Republic consisting of 15 ministers, including the prime minister. The Prime Minister is the head of the Government and is presented by the president to parliament and is authorised by parliament (OECD 2001a). Estonia has a central government administration and a local independent, self-governing administration. Central government is represented through a single regional level. The local administration is also represented by a single level. The Estonian National Strategy (NES) was approved by Parliament in March 1997. The NES specifies the trends and priority goals of environmental management and protection in a new political and economic situation and sets the main short-term and long-term tasks to be achieved by the years 2000 and 2010, respectively. The strategy follows the main international environmental documents, including the principal requirements of integration of the associated countries of Central and Eastern Europe with the European Union (EU) (the White Paper) (Ruut 2001). The NES was followed by the National Environmental Action Plan (NEAP), which was adopted by the Government in May 1998. The NEAP's main principle is the integration of environmental concerns into other sectors. It also sets environmental priority issues for which specific activities are foreseen, guides the planning of environmental expenditures in state budget and other budgets, starts process of periodic assessment and revision of environmental priorities and actions, assists consensus building for environmental decision-making and co-ordinating participation of different stakeholders. The Estonian National Commission on Sustainable Development was established in 1996 and operates under the leadership of the Prime Minister.

3.4.1 The Ministry of the Environment

The Ministry of the Environment is led by the Minister of the Environment. The chief executive of the Ministry was the Secretary General who was in charge of managing the work of the sub-units of the Ministry and co-ordinating the activities of the institutions governed by the Ministry. Directly subjected to the Secretary General were four Deputy General Secretaries, each with special fields of competence. The Ministry of the Environment was responsible for the protection of nature and environment, land use, the co-ordination of the elaboration of regional plans, managing the use, protection and accounting of the natural resources, as well as surveillance over the use of environmentally hazardous compounds. The Ministry organised environmental monitoring, meteorological, geological, constructional, and geodetic surveys and research in natural history. It arranged environmental impact assessments and co-ordinated international relations in environmental matters. The Ministry formulated national policies in its field of activities and prepared the bills of respective legal acts. There were also 15 land boards, also known as cadastral offices. The principles of Estonia's post-Soviet environmental policy included the acts on Nature Conservation, Forest, Protected Natural Objects, Protection of Marine and Freshwater Coasts, Shores and Banks, Pollution Charges, Water, Earth's Crust, Packaging, Planning and Building, Hunting Management, Fishing, Land Amelioration, Sustainable Development and Health Protection.

3.4.2 County environmental departments

The county environmental departments submitted relevant reports to the Ministry and the County Governor. They also administrated protected natural objects, selected key habitats and proposed habitat protection contracts. In the area of water management, the County environmental departments were given the task of managing the protection of ground water quality and resources including bored wells and important springs. They were expected to maintain registers of still and flowing water bodies, to draft programmes for the renewal and protection of fish resources, propose fishing gear prohibitions, fishing limitations and desired fishing permit fees, and manage the collection of fishing and fisheries renewal data. They also had an important administrative purpose as they were in charge of information gathering and reporting in the field of the environment and nature use, and were expected to carry out activities in the field of public mobilisation, environmental education and publicity.

3.4.3 Local government in Estonia

Over the last decade, there has been a transfer in responsibility for the environment; however, this has not been accompanied by a clear definition and division of roles. Central government continues to undertake environment-related actions through environment or transport ministries and their own regional bodies, such as County Environmental Departments. These duties often include inspections, award of permits, environment monitoring, data collection, and dissemination. Local governments are left only with clear responsibility for public utility services like wastewater treatment, drinking water supply, and waste management. Local government tend to be responsible for water supply and sewerage, heating supply, waste collection and care for landfill sites, and land use planning. In the transport sector, responsibility frequently tends to remain with the transport ministry and its associated bodies, however, local government do take charge of maintenance of municipal and rural roads and streets, street signs and lighting, investments, operation and financing. The unclear definition of roles has also been accompanied by poor financial support. Most local government administrations today rely on the following sources to fund their activities: state revenues including income tax allocation, state budget subsidies, local fees, fines and taxes, loans, donations, and grants.

The basic provisions concerning local government level in Estonia are laid down by the Constitution of 1992. The Constitution recognises the principle of local government and states that all local issues shall be resolved and managed by local authorities, which shall operate independently pursuant to law. Until October 1993 self-government existed both at local and regional level. Since then the counties ceased to exist as self-government authorities and Estonia had a one-tier local government system. In accordance with the Local Government Organisation Act and the Territory of Estonia Administrative Division Act, Estonia was during the early 2000s divided into 15 counties and 247 local government units all with the same status (OECD 2001b). Among the 247 local governments, there were 42 towns and 205 rural municipalities (OECD 2001b). The size of local government units varied considerably. The towns with the largest population were Tallinn and Tartu with 411,594 and 120,000 inhabitants, respectively. The smallest municipality was Ruhnu, a rural municipality on the island of Ruhnu, with 72 inhabitants.

Towns and rural municipalities were the units of local government and they were equal in their legal status. They could be divided into municipal districts (with a limited right to self-government) in order to carry out public services. All municipalities were part of a county. The county governments were units of state administration and were financed from the state budget. County governments had supervisory and advisory functions concerning self-government and County administration served as the administrative apparatus of the county governor. The county governor headed the county government and was appointed by – and subordinated to – the state government. She/he represented the central government and implemented its policy at the regional level. The Government of the Republic appointed the county governor to office after approval by local government representatives.

As there was no regional government level in Estonia, the co-operation of municipalities within a county was of great importance to municipalities both in their relations with central authorities and in co-ordinating their own activities. Local authorities had a constitutional right to form associations or establish joint institutions. The formation of association was specified in the Local Government Organisation Act. Municipal governments had also chosen to establish a representative association, which in some cases actively served to encourage municipal economic and cultural development. Representatives of these organisations were also entitled to participate in central government cabinet hearings and parliamentary discussions, and could also participate in drafting state budget allocations, these bodies were also likely to serve as reference points for EU related information and project opportunities. Assisting local governments in preparing for EU membership was considered a political priority.

Associations of local authorities within counties carried out tasks set by the central government as well as by local authorities. Tasks set for the associations by central authorities were the evaluation of higher officials of administrative agencies in the municipalities of a county and the officials of the bureau of the local government association in the county, except the head of the bureau, concordance of national planning policy statements and making a proposal to the county governor for the distribution of the regional support fund between the municipalities of a county. Tasks set by member municipalities ranged over a variety of fields. The associations exercised construction supervision, co-ordinated public transport within a county, participated in the drafting of county development plans, were engaged in the development of the use of information technology, provided technical and legal advice, promoted tourism, arranged inner-county cultural and sports events, participated in educational projects, etc. The advantage of inter-municipal co-operation was having the flexibility to co-operate in any field and through several types of co-operation (buying service from another municipality, hiring a common expert or specialist, founding profit and non-profit associations). Inter-municipal co-operation could be evaluated and adjusted within a reasonable length of time after changes in the external framework, for example, new laws, new technology, new actors, and new demands. The disadvantages of inter-municipal co-operation were connected to paying for the services provided and, occasionally, the lack of willingness to co-operate, even in the cases where co-operation was more efficient than a single local government unit providing the service.

In all the 15 counties, there was an association of local authorities. In 13 counties, the association united all of the municipalities of the county. In two counties, Pärnu and Harju, some municipalities were not members of the association. In addition, the municipalities of Pärnu and Harju were divided into two regional associations. There were three national associations of local authorities in Estonia. They were the Association of Estonian cities (established in 1920), the Association of Estonian Rural Municipalities (established in 1921) and the Union of Estonian Associations of Local Authorities (established in 1993). The basic functions of the national associations were the representation and protection of local government interests and participation in the legislative process concerning local authorities. In 1994, a joint body of the three association was formed, the Co-operation Assembly of Associations of Local Authorities, in order to conduct negotiations with central government and provide co-ordination between member associations.

There was a need to reduce the number of local governments; many functions were relatively more expensive in smaller local governments. However, this did not necessarily mean that bigger ones were more effective. At the same time, there was an extensive decentralisation from central government to the local level of government. The increase of administrative tasks and services placed a strain on local finances as many municipalities were comparatively small. Fifty percent of the 247 municipalities had less than 2000 inhabitants. They therefore often lacked the sufficient economic, administrative and personnel capacity to carry out the ever growing amount of tasks efficiently. The opportunity for municipal governments to make inspections and ensure enforcement in the environmental field was limited because this tended to be handled by an environment board or inspectorate. Furthermore, municipal administrations rarely dedicated a department or individual with responsibility for environmental issues, in part also reflecting the low level of importance accorded to environmental issues. In many cases, these persons were employed on a part-time basis only. Potential co-operation was further hindered by a shortage of funds and limited English language skills.

3.4.4 Environmental problems in Estonia

From the Soviet period, Estonia inherited several serious environmental problems. The main air polluters in Estonia were two thermal power plants in the north-eastern part of the country. The socialist intensive agriculture, which concentrated on the needs of the Soviet market, in a combination with the unbalanced use of mineral fertilisers, mainly nitrogen fertilisers, also caused contamination of surface and ground waters and soil. There were four main causes of environmental problems. The first was the use of out-of-date technologies consuming large volumes of raw material and generating large quantities of waste. The second was the low level of public awareness and a consumption-oriented mode of thinking. The third was the underdeveloped environmental–technical infrastructure. Finally, insufficiency of financial resources and management instruments create major problems (2001). Also, while Estonia had made progress with its transposition of framework Directives, there was, however, also a considerable need to improve administrative capacity to implement and enforce legislation. This was especially true at the local level (2001).

There were at the time seven priority environmental problems in Estonia. Firstly, there was the pollution of ambient and urban air, which had a negative impact on human health, ecosystems, and buildings. Secondly, there was past pollution caused by industrial, agricultural, and military activities that posed a threat to ground and surface water quality. Thirdly, there was a decrease in water quantity and quality due to irrational use of ground water resources. A fourth problem was the irrational use, pollution and eutrophication of surface water bodies, deterioration of aquatic ecosystems, including a decrease in reproduction and deterioration of the quality of fish stock. Fifthly, there was the increase in environmental pollution caused by waste disposal and in areas contaminated by waste, inappropriate waste and hazardous waste management. Threats to biological and landscape diversity as a result of economic activities and land reform were the sixth problem. Finally, there was the problem of insufficient correspondence of built environment to environmental and health principles (2001).

3.4.5 The Common Agricultural Policy (CAP) and its role in the scenarios

The CAP was, and still is, one of the EU's most debated policies. Not only against the background of the enlargement process, but also in itself due to the complications by bureaucratic and political communities at both national and European-levels. At the time of the MANTRA-East project, it was obvious that the CAP once again was in need of reform. The EU member states all agreed that in order to cope with the economic pressure imposed by the enlargement fundamental change was necessary. Differences of opinions began when it came to exactly what the reforms should entail and when the reforms should take place. There were several difficulties to be taken into consideration: one was most certainly the double loyalty of the decision-makers, loyalty to their nations and loyalty to the EU as an organisation. The most obvious difficulty was by far the competing national interests of the 15-member states in 2003. It was difficult to see how the net contributor in terms of financing the CAP (Germany) could share national interests with the net beneficiary (France). As enlargement process took place, it was reasonable to assume that Germany would want to scale down the CAP budget and that France would prefer a reform where the French farmers would receive, if not more money, at the very least the same amount. These two preferences were obviously not compatible. The main question was then, what would the future CAP look like? This was of course important for the construction of the scenarios and the modelling processes described in the coming chapters. There were several ideas of how financial support could be structured under a new CAP system, but perhaps the three most relevant were:

(1) a transitional period;
(2) a system where the present member states would receive the same amount of agricultural support and the new member states would receive less; and/or
(3) a development and reformation of the accession countries' agricultural sector, where the relative and real size of their agriculture would be scaled down.

The future of the CAP and the timing of reforms were difficult to predict. The Commission had over the years experienced extreme difficulties in reforming the CAP and in moving it away from its original course. However, some changes had also occurred surprisingly fast. In 1992, agricultural policy changed sharply after a relatively short debate, in the Uruguay Round the new commitments were adopted with almost no resistance and at the beginning of the 2000s the CAP had undergone a further change in preparing for the enlargement. So, even though it might have seemed as if the CAP was a slow-changing ever-continuing policy, changes can (and had) occur quickly under the right conditions. Therefore, it was necessary to formulate a long-term perspective of the developments of the CAP.

As the CAP was so central to environmental management in Estonia this section presents an account of the CAP, dealing with its past and present, in order to comment on its future. The section attempts to analyse the early 2000 debate over the future developments and changes of the CAP, changes thought necessary as the enlargement from 15 to perhaps as many as 27 members of the EU took place. In order to do so, it is necessary to understand the different perspectives and positions of the pre-2004 15-member states, as well as the policy-making process of the EU. There were three major assumptions underlying this section, namely: that there were, at least, three possible directions for the CAP reform to take:

(a) a transitional period for a number of years where the pre-2004 and new member states would adjust in order to work out a long-term sustainable solution for the CAP;
(b) a system where the pre-2004 member states would receive the same amount of agricultural support and the new member states would receive less; and
(c) a development and reformation of the new EU countries' agricultural sector, where the relative and real size of their agriculture would be scaled down.

In this section these three directions are examined further by identifying the member states' individual national preferences.

The work in the following section is based on information gathered through studies of the websites of national and European newspapers. There were a large number of articles written on agriculture and enlargement, and by simply using the two terms as search criteria, the results numbered over 78,000 hits. The articles selected were the most recent ones (in 2002), concerning the CAP and reforms in the 1990s with focus on debates over the latest years. The section begins with a background orientation of the CAP, its history and its main principles and objectives. A brief section deals with the decision-making and policy style of the union, followed by a reform overview focusing on the MacSharry plan and Agenda 2000. The main part of the section on the CAP concentrates on the debates in the member states and European institutions that occurred at the time of the MANTRA-East project.

3.5 BACKGROUND OF THE CAP

The aim of the CAP is to increase the productivity of European agriculture, ensure European farmers a reasonable standard of living, stabilise farm produce markets and guarantee a stable food supply at fair prices for European consumers. The means have, since the inception in 1962, been agricultural price supports, direct payments to farmers and supply control. The reforms of the CAP enacted in 1992 and under Agenda 2000 moved the EU in the direction of providing support through direct payments to farmers/producers, rather than through price support policies (Hix 1999).

When the CAP was first set up the main objective was to ensure a self-sufficient Europe in agricultural goods. The two world wars during the 20th-century had shown just how vulnerable Europe as a society/continent was in terms of not being self-sufficient. As production stabilised and Europe became a net exporter of agricultural products the price of support mechanisms led to several difficulties. Guaranteed prices stimulated production but eventually led to overproduction, over-use of herbicides, pesticides and artificial fertilisers. Over-intensive land use resulted in environmental problems. Import quotas and levies prevented global-free trade in agricultural goods and export subsidies contributed to global development problems as they depressed world prices and distorted agriculture markets in the Third World. In this regard, the EU's agriculture has become more and more subject to the pressures of GATT agreements, governed by WTO rules (Brennan 2000).

Agricultural trans-national co-operation proved effective and agricultural production increased from the inception of the CAP in 1962. The shift in Europe from vulnerability to self-sufficiency, however, made CAP reforms necessary as overproduction and other above mentioned problems emerged. From embracing policies to stimulate production, overproduction had to be limited. The result, due to the incrementalistic decision/policy-making style of the EU, is a system with counterproductive policies, including policies to stimulate production in some geographical areas, and policies to reduce production in others. From the public's point of view these policies have come to turn opinion against the CAP, regarding agricultural co-operation as something often seen as a terrible waste. Agricultural products are left in countries outside the geographical hegemony of the Union and sold to consumers at dumped prices, thereby distorting national markets. Also, the fact that farmers are paid not to grow certain crops on some of their farms, or not to grow anything at all, has led to the public's decreasing support of the CAP.

Article 33 (ex. 39) of The Treaty of Rome sets out the CAP objectives as follows:

(1) to ensure agricultural productivity, by promoting technical progress and ensuring rational development of agricultural production and the optimum utilisation of the factors of production, particularly labour (stabilisation principle);
(2) to ensure a fair standard of living for the agricultural community, in particular by increasing the individual earnings of persons engaged in agriculture (redistribution principle);
(3) to stabilise markets (stabilisation principle);
(4) to ensure the availability of supply (allocation principle); and
(5) to ensure that supplies reach consumers at reasonable prices (allocation principle).

The key principles underpinning the objectives of the CAP, the Single Market, Community Preference and Financial Solidarity were laid down at the Stresa Conference in 1958, and are still the main principles upon which the CAP is based. Firstly, the Single Market guarantees free flow of agricultural commodities between the community's member states ensured by a general prohibition on levies or subsidies by national governments, common euro prices and harmonisation of technical regulations. Secondly, Community Preference guarantees priority for the community's agricultural products over imports in the internal market, assured through protection from imports. Thirdly, Financial Solidarity means common financing of agricultural programmes from a common fund financed through member states assessments and revenues from certain levies (Grant *et al.* 2000; Wallace and Wallace 2000). In 1962, the member states agreed to put the principles into practice. The result is a complex system of indirect income support for farmers, a support paid for by the European citizens; through the EU budget, that is, by European taxpayers; and by European consumers through the extra prices charged on imported agricultural goods of indirect income support for farmers (Hix 1999).

3.5.1 The reforms

From the time of the inception of the CAP agriculture has developed to a highly technological industry, employing only about 4–5% of the EU population. As the incitements to produce more agricultural products have in fact led to overproduction, reforms have been necessary. During the first 30 years of implementation of the CAP, surprisingly little change occurred, with the notable exception of milk quotas introduced in 1984. There were of course some changes, but none of which would constitute a 'reform' (Moyer and Josling 1990). The first two changes during the 1970s and 1980s, included:

- modified price policy,
- restricted market intervention,
- regulated output,
- introduction of a system to control spending on agriculture.

Several measures to stimulate development in rural areas where added to the change packages. These measures included training for farmers, financial assistance for the youngest farmers, investment in modernising farm holdings, and special programme to assist less favoured areas and mountain areas.

The current CAP system is the result of mainly two reform packages, the MacSharry plan adopted in 1992 and Agenda 2000 adopted in 1999. The two reforms began the process of shifting farm support from prices to direct payments. Even though the original principles of the CAP and their implementation were successful during the first decade, leading to a stable growth in agricultural output, in turn leading to self-sufficiency in food in the EEC-6 accompanied by low consumer prices, a growing surplus soon became a problem, creating public stocks and a surge in agricultural spending. The CAP was in need of reform. The first major change in the instrumentality of the CAP was the MacSharry plan in 1992, when a more radical overhaul of the CAP was called for as it became apparent that agricultural products needed to be more competitive in terms of quality and price, as consumers expectations were growing. The 1992 MacSharry plan (named after the Irish Agricultural Commissioner) contained four main policy changes: price cuts, direct income support, a set-aside scheme, and accompanying measures. The price cuts applied to certain sectors, the guarantee price for cereals and beef were reduced by a nominal 29% and 15%, respectively, over a 3-year period, and aimed at restoring market balance. The price levels of agricultural products were adjusted to levels much closer to world market levels. Support prices were reduced and direct compensatory payments for farmers introduced to compensate for price the reductions. The set-aside scheme means that farmers are paid not to farm, farmers in specific regions are paid to leave their land fallow in addition to producing agricultural products leading to surpluses that would have to be bought by the EU. The accompanying measures include the introduction of aid programmes to promote rural development, environmental friendly farming, early retirement of farmers, and the replacement of agricultural land with forests (Hix 1999). In sum, the importance of the MacSharry reform was to shift emphasis away from farm incomes. Additional changes were implemented in 1995 as part of the EU's commitments in the Uruguay Round agreement on agriculture, anticipated as well as facilitated by the MacSharry reforms. The changes covered all farm products and required a 20% reduction in the aggregate level of domestic support for agriculture over a six-year period and included conversion of variable import levies to tariffs (Josling and Babinard 2000).

At the time of the MANTRA-East project it was thought that as the enlargement of the union took place, the MacSharry reforms would not be sufficient. In the pre-2004 candidate countries agriculture was an important sector, but not nearly as technologically developed as in the existing EU. In the applicant countries agriculture was relatively more important than in the 15-member state EU. The Central and Eastern European Countries (CEEC) had a generally higher agricultural labour force compared to the pre-2004 member states. A larger share of the population was employed in and dependent on agricultural markets in the CEECs than in the union, on average 22% (9.5 million workers) of the workforce in the CEECs is employed in agriculture compared to an EU average of 5% (8.2 million). In Poland, considered a 'problem' country when it came to agriculture and enlargement, the agricultural employment rate was relatively even higher with 25% of the total employment and an agricultural land area amounting to 58.9% of the total land area. However, the agricultural share of the GDP was only 3.8%. Moreover, surveys showed that farmers in Poland were sceptical to the union (Economist 2001). In total, with enlargement, agriculture in the EU would increase from 60 to 200 million hectares and the agricultural labour force would at least double.

At the end of the 1990s, a long-term budget that was to take the enlargement into consideration was needed and the Commission came up with a blueprint for those aspects of EU policy most affected. The Agenda 2000, first introduced by the EU Commission in July 1997, was based on the 1992 reforms, adopted in a somewhat watered out version in March 1999 and implementation began in 2000. The March 11th Agreement was further weakened after French insistence in particular by the Berlin Summit 2 weeks later (George and Bache 2000). Behind the reforms lay the fact that since 1992, new internal and external challenges had arisen. These included the unequal distribution of support between regions and producers, leading to a decline in agriculture in some regions and overly intensive farming in others, generating pollution, animal diseases, and poorer food safety. Also, the current levels of CAP prices were too high for the EU to meet international undertakings and to take advantage of the predicted world market expansion in agriculture. This generated the risk that surpluses would reappear accompanied by an intolerable budget cost (Hix 1999). The main aims of the proposal in 1998 were to reduce the level of commodity prices, improve the balance of support to focus on those areas most disadvantaged, improve public perception and acceptance of the CAP and focus on ecologically sustainable agriculture rather then maximal production. In sum, the Agenda 2000 reform set out to deepen and widen the 1992 reform by replacing price support measures by direct payments to farmers (65% of the subsidies should go straight to farmers), accompanied by a consistent rural policy.

The negotiations on agricultural policy demonstrated that they continued to be one of the more difficult issues in the enlargement negotiations of the EU. One of the major challenges was the development of policies to modernise and rationalise agriculture in the CEECs, which had suffered from under-investment and lack of infra-structural support, and at the same time restructure the EU itself and its agricultural policies. The result of enlargement was an expected increase of 50% of agricultural land in the EU and a 100% increase in agricultural labour. Without CAP reform, away from price support, enlargement was seen as likely to drain the EU budget. The EU budget funds were, in the long-term, allocated for agricultural policy in the new member countries for the period until 2006, but the funds did not cover the costs of direct subsidies for farmers, and many candidate countries demanded such direct subsidies. A key point was that when the next multi-annual framework for the period 2007–2012 was to be decided in 2004–2005, the CEECs took part in the negotiations. The CEECs argued that present spending rules should be applied to them, whereas the former net beneficiaries argued that they could not be expected to accept less, and that their populations cannot and would not accept an increase in transfers into the EU system.

A second main issue was that of competition and how EU farmers would be affected by competition from the new member states. Generally, due to under-investment and lack of infra-structural support and market economy, farms in the CEECs were low productive, small, and oriented towards supporting families, not towards competition on the common market. Support was therefore provided for structural change through the Special Accession Programme for Agriculture and Rural Development (SAPARD). The SAPARD programme was criticised in a report by Willy Görlach for the European Parliament's agriculture committee. Görlach claimed the scheme to be '*poorly organised*'. He added that it is '*regrettable that a project launched with such great expectations as SAPARD is for the most part still not up and running*'. The background of the statement is that in 2001, only half of the potential beneficiaries – Bulgaria, Estonia, Slovenia, Lithuania, and Latvia – had received any of the allocation. The report found that out of the 530 billion euro per year earmarked for the programme for 2000–2006 only 30 billion euro flowed into those countries between the start of the (SAPARD) in 1999 and the end of 2001.

3.5.2 Political control and decision-making

The EU policy-process has been conceptualised as at structure of multi-level governance, characterised both the decision-making within the union and the relationship between EU institutions and institutions in the member states (Pierre 2000; Richardsson 2000; Wallace and Wallace 2000). The concept of multi-level governance implies that there is a web of network-like interactions between a wide range of different actors, such as state and sub-state, public and private, national, transnational and supranational, NGO's etceteras, in turn suggesting that the national state no longer monopolises the policy-making processes (Payne 2000). In such an 'anarchic' system, where exogenous and endogenous forces have made it less possible for the state to steer in traditional ways, alternative systems of governance have emerged. This means that since no single actor alone controls the policy-making process, policy-outcome is characterised by a high-level of uncertainty undermining stable patterns at national levels. Richardson suggests that governance at the

EU level can best be described as uncertain agendas, shifting networks and complex coalitions (Richardsson 2000), governance theories explicitly involve other actors than states. Pierre and Peters (Pierre and Peters 2000) claim that contrary to much of what has been written about the 'decline of the state', that the emergence of governance could actually increase rather than decrease state and public control over society. Even though governance relates to changing patterns in terms of decision-making and power relations between state/community the state remains the centre of political power. Governance is a process where the leading role is held by the state; the state defines goals, sets priorities, co-ordinates, undermines or reinforces the power of actors by choosing whom to co-operate with, etc. and the state remains the best possible tool to force implementation (Pierre 2000).

The agricultural policy-process in the EU is characterised by the influence of a wide range of non-institutional members and interests. Farming lobbyists and farming interests have the possibility to steer policy in certain directions, especially in the early stages of the policy-process. Agricultural policy-making is no exception to the multi-level governance style of EU policy-making.

Agriculture was one of two EU sectors where there was a common policy, that is, agricultural policy was made by a supranational authority and subject to agreement by the member states agricultural ministers and sometimes require approval by the EP (the other was transport). However, there were a number of committees ensuring the prominence of national interest in this apparently supranational policy-process (Rieger 2000) and agriculture policy was made by agriculture ministers, agricultural officials in the Commission and European-level farming interests. The EP's role was limited. Agriculture ministers were usually recruited from political parties supported by farmers and representing rural regions, and the work of the Agriculture Council was supported by the Special Committee of Agriculture (SPA), staffed by officials from national agriculture ministries, rather than the usual Committee of Permanent Representatives (COREPER), staffed by career diplomats. This meant that the agriculture policy-process was strongly influenced by national interests, and proposals to re-nationalise decision-making were expected to further bring it under national control. Agricultural interests were further protected by the way the Commission managed the CAP; the agriculture commissioner had always been recruited from a farming state, and usually came from a political party supported by farmers, or at least from a party with close links to farming interests. Moreover, on the national level farmers were more or less ensured a central policy-making role due to the corporatist relationship between farmers' associations and agriculture ministries. The European-level equivalent to this relationship is found in the Confederation of Professional Agricultural Organisations – the most well-resourced and well-staffed of the supranational sectoral associations. All this adds up to what Hix refers to as 'the CAP iron triangle' (Hix 1999, p. 253) meaning that each element in the decision- and policy-making process of agricultural policy has vested interests to defend the interests of the other. The iron triangle is further reinforced by the consumer's lack of incentive to break the triangle since the cost of mobilising and forming an anti-CAP campaign is higher than the cost paid by each individual consumer/taxpayer. The role of the Commission and the influence of national influence mean that the

CAP is something like a halfway house between the nation state and the truly supranational. There are, however, two major developments that have undermined the CAP iron triangle:

(1) Social, economic, and political changes in Europe have reduced the power of agricultural interests. The number of farmers is constantly declining and so is the status of agriculture in the national economy;
(2) The development of external global trade pressures has created incentives for CAP reform, starting with the GATT and Uruguay Round where it was stated that without a reform/cut in subsidies to farmers an agreement on global trade liberalisation could not be achieved.

3.5.3 The future of the CAP

It was believed that the direction of the future CAP depended on several factors; the enlargement was one of the primary two, budgetary constraints the second (Josling and Babinard, 2000). In addition, pressure to reform also came from environmental and consumer concerns, as well as the external world, mainly the WTO and the USA. Moreover, the structure of multi-level governance undermines stable patterns and creates uncertainty of policy-outcome. The policy-process of the EU is also far too complex and incremental to steer policy in a specific direction in a pre-implementation phase. This puts constraints on the possibilities of decision-makers. To begin with, attempting to implement policies without the support of NGOs, scientific and other networks and more especially without the support of farmer's organisations such as the Confederation of Professional Agricultural Organisations is not likely to succeed. Secondly, a policy in a multi-level governance system is constantly under the influence of various actors, both at local, regional, national European level, the actors being both governmental and non-governmental, implying that the policy-outcome is not likely to be equivalent to the policy-intention. This adds up to the following: whatever analysis one might make drawing upon the various statements and debates over the future of the CAP, the future is more than likely to differ from future predicted. This is one of the reasons that we talk about decision-making under conditions of uncertainty. However, the future of the CAP (with some reservation) was expected to probably take one of the three directions suggested in the introduction. These three suggestions of possible reform directions of the CAP were:

(a) A transitional period for a number of years during which the pre-2004 and new member states would adjust in order to work out a long term sustainable solution for the CAP.
(b) A system of uneven distribution between new and pre-2004 member states where the present member states would receive the same amount of agricultural support and the new member states would receive less.
(c) A reformation of the accession counties' agriculture where the relative and real size of their agriculture would be scaled down.

Those countries that were net beneficiaries of the CAP subsidies were expected to probably prefer the second alternative, a system of uneven distribution. This would allow France and Spain to keep the same amount of subsidies, and at the same time not fear that the enlargement would scale down their share of support. There were some tensions between the French and the Commission on this point, both were likely to prefer a transitional period but the Commission saw this as a way of dealing with the enlargement first, the CAP reform later. To the French, a transitional period meant that the assumed negative financial consequences of enlargement became a future problem. One has to remember though, that the French population's opinions of farm support seem to have shifted recently, a shift away from 'farmer supportive' towards a more CAP reform friendly attitude. In Germany, however, being the net contributor, this was not a favourable solution. The outspoken German preference of the time was re-nationalisation of farm subsidies and the second alternative did not create any incentive for the beneficiaries to take on a national financial responsibility for agriculture since the money would keep on coming from the Union to the same extent as before. Whatever reform was preferred by the Germans at the time, it included some possibility of realising the re-nationalisation plan, or at least some system where the idea of re-nationalisation was not totally out of the picture. Second, to hang on to a system of uneven distribution was not in tune with the German idea of extended solidarity. On the contrary, a system of uneven distribution would mean grading the members of the union into separate divisions of solidarity, the higher division being the old and pre-2004 members who would enjoy a kind of whole hearted solidarity commitment in terms of financial support whereas the new members would enjoy a half hearted solidarity commitment. Third, the German view of solidarity can be interpreted as the opinion that the wealth of countries such as France, Portugal, and Spain is relatively high compared to that of Estonia, Latvia, or Poland, meaning that 'solidarity' per definition has to include scaling down the French and Spanish shares. Drawing upon the above three arguments, a system of uneven distribution was expected to meet opposition from Germany.

Of the three alternatives, the third was the one least commented on, even though this might have been the easiest way out and also the only long-term sustainable solution. Without claiming that agriculture has lost its importance, it would only be fair to say that modern technology has made it possible to reduce the agricultural labour force and the number of farms without ending up with a decrease in production. On the other hand, a decrease in production would, all other things being equal, solve some of the financial problems caused by the Commission being obliged to buy agricultural surpluses products to keep price and demand stable. The EU was, despite previous reforms, dealing with problems of overproduction and was therefore not in need of policies promoting production, but rather policies limiting it. It was seen as quite possible to impose strict production controls on the CEECs as part of their entry condition. As a step in the direction of making way for the CEECs, reduced EU subsidised exports were thought to be a step in the right direction. And also, the very idea of the CAP was to ensure a self-sufficient Europe. This idea is no longer the underlying principle, at least not in the same extent as it was in the 1960s. Instead the principles of the 21st-century seem to take other aspects into consideration, aspects such as free trade, globalisation, helping (or simply giving back to) the third world, sound animal treatment, sustainable environmental development, and 'safe food'. In a system where these aspects are to be considered, a system such as the CAP is seriously outmoded. To reform the CEEC's agriculture was seen as perhaps the only sustainable solution. The transitional period and the uneven distribution systems both represented temporary solutions. They might also both have been acceptable as such, but they lacked a long-term perspective making them rather inappropriate as main principles in an enlarging Union aiming at stability and long-term prosper. To reform the CEEC's agriculture was seen as a solution that did not directly oppose the national interests of the Union's members. It could have been in tune with a more ecological agriculture (Great Britain), it could have scaled down the expected budgetary constraints (Germany) and thereby allowed for the pre-2004 members to, at least temporarily, receive the same amount of subsidies (France, Spain). However, this alternative needed to be accompanied by an all over cut in agriculture/CAP expenditure. This is where the tensions and future consensus between Germany and France became vital.

In some aspects, the future of the CAP could be better predicted by using an analysis of power. In the end, the reform of the CAP did not really depend on rational arguments but on relative *national* power. The real question then came down to: How relatively powerful are the different national and European institutions? Was the Commission powerful enough to push forward its CAP plan or was the German or French sphere of influence greater? Or was someone else more powerful? The brief presentation of governance theory presented above suggests that the power of the Union remains in the hands of national governments, without whose consent policies will not be decided upon or implemented. The reforms of the CAP agreed upon in 2003 are not treated here as they were not incorporated into the scenarios that were produced in the MANTRA-East project. Many would also claim that the reforms will be of limited effect and that major changes in the CAP are still necessary.

3.6 PROBLEMS AND CO-OPERATION

On the ground, the opinions of environmental managers in Estonia and the Russian Federation expressed in the interviews show that environmental administrators found it difficult to describe the decision-making processes, both of their own countries, and especially of their counterparts across the border. They also had difficulty in describing the central driving ideas behind environmental management. This indicates a lack of institutional transparency, especially in Russia. Hierarchical decision-making processes were common, although while it is most usual that the heads of department approve all decisions, the specialists also had a reasonable degree of independence to organise their work. The importance of national and European legislation is notable, and the dominating influence on environmental management is seen to be the regulatory legislative structures. The concept of subsidiarity seemed not to include local levels to a significant extent, and at the local level the decision-making process was seen by many (but not all) of the interviewees as top-down. The problems described by the environmental managers were in some ways similar. They ranged from lack of resources, lack of co-ordination of tasks, to the influence of party politics in environmental administration. As transboundary relations were a central focus of the MANTRA-East project, the interviewees were also asked about co-operation. Co-operation between local authorities was considered positive, and a number of examples of successful co-operation were described. Relations between the municipal and county authorities were not, however, described as a prioritised issue by the local authorities, although co-operation with the County Environmental Department was good. Many of the interviewees felt that local-state relations were more important. Here it was possible to suspect that the inter-organisational competition that was apparent on the Russian side also exists on the Estonian side of the border. Officials at the county level also felt that communication with the central level, in the form of the Ministry of the Environment, were too hierarchical. As in Russia, water pollution was considered the major problem on the Estonian side, although it is interesting to note that the majority of the interviewees did not consider agriculture the major problem in this respect. The reasons for this discrepancy was that many of the interviewees felt that the decrease in agricultural production, as well as the decrease in the use of artificial fertilisers, had led to a decrease in the magnitude of the problems of diffuse pollution from farming land. Research within the MANTRA-East project has shown, however, that this may not be the case, and that residual effects are significant. Forestry, although also mentioned as a problem, was not either considered a major issue.

REFERENCES

http://www.ecouncil.ac.cr/rio/national/reports/europe/estonia.htm (2001).

Barry, J. and Frankland, G. (2002). *International Encyclopedia of Environmental Politics*. Routledge, London.

Bremmer, I. and Taras, R. (eds) (1997). *New States, New Politics: Building the Post-Soviet Nations*. Cambridge University Press, Cambridge.

Brennan, I. (2000). The challenge of the East. In: *Contemporary Europe*, R. Sakwa and A. Stevens (eds). Macmillan Press Ltd, London.

Brown, A. (2001). *Contemporary Russian Politics*. Oxford University Press, New York.

Budarin, V., Sedova, A., *et al.* (2001). Statute of the RF Ministry of Natural Resources. *Water Resource Management in Russia: Lake Peipsi Basin Case*. CTC, Tartu.

Crotty, J. (2002). Economic transition and pollution control in the Russian Federation: beyond pollution intensification. *Europe–Asia Studies* **54**(2), 299–316.

Economist, T. (2001).

Evangelista, M. (2000). Russia's path to a new regional policy. PONARS Policy Memo No. 157, 2002.

Gelman, V. (2001). The rise and fall of federal reform in Russia. PONARS Policy Memo No. 238, 2002.

George, S. and Bache, I. (2000). *Politics in the European Union*. Oxford University Press, Oxford.

Glenny, M. (1993). *The Rebirth of History*. Penguin, London.

Grant, W., Matthews, D., *et al.* (2000). *The Effectiveness of European Union Environmental Policy*. Macmillan Press Ltd, New York.

Hix, S. (1999). *The Political System of the European Union*. Macmillan Press Ltd, London.

Ivanova, M. (2001). Environmental crime and punishment in Russia: law as reason for breach. *Conference on the Human Dimensions of Global Environmental Change*, Berlin.

Josling, T. and Babinard, J. (2000). The future of the CAP and prospects for change. *The Policy Environment for Agri-food Competitiveness*.

Kelley, D. (1999). *Politics in Russia and the Successor States*. Harcourt Brace College Publishers, Fort Worth.

Klyamkin, I. and Shevtsova, L. (1999). *This Omnipotent and Impotent Government: The Evolution of the Political System in Post-Communist Russia*. Gendalf Publishers, Carnegie Moscow Center, Moscow.

Kourlindskaia, G. and Kachanova, E.A. (2002). Local government in the Russian Federation. In: *Developing New Rules in the Old Environment*, I. Munteanu and V. Popa (eds). Local Government and Public Reform Initiative, Budapest.

Kuzmin, A. and Kachanova, E.A. (2002). Local government budgeting: Russia. In: *Local Government Budgeting*, M. Högye (ed.). Local Government and Public Reform Initiative, Budapest.

Massa, I. and Tynkkynen, V.P. (2001). *The Struggle for Russian Environmental Policy*. Kikimora Publications, Helsinki.

Melvin, N. (1995). *Russians Beyond Russia: The Politics of National Identity*. Royal Institute of International Affairs, London.

Moyer, H. and Josling (1990). *Agricultural Policy Reform: Politics and Process in the EC and the US*. Hemel Hemstead, Harvester Wheatsheaf.

OECD (2001a). Fiscal design across levels of government year 2000 surveys. Country Report, Estonia, 8.

OECD (2001b). Fiscal design across levels of government year 2000 surveys. Country Report, Estonia, 2002.

Payne, A. (2000). Globalization and modes of regionalist governance. In: *Debating Governance*, J. Pierre (ed.). Oxford University Press, Oxford, pp. 201–218.

Peterson, D.J. and Bielke, E. (2001). The reorganization of Russia's environmental bureaucracy: implications and prospects. *Post-Soviet Geography and Economics* **42**(1), 65–76.

Petrov, N. (2001). Policization versus democratization: 20 months of Putin's 'federal' reform. PONARS Policy Memo No. 241, 2002.

Pierre, J. (ed.) (2000). *Debating governance: Authority, Steering, and Democracy*. Oxford University Press, Oxford.

Pierre, J. and Peters, B.G. (2000). *Governance, Politics and the State*. Macmillan Press Ltd, Houndmills, Basingstoke and London.

Richardsson, J. (2000). Government, interest groups and policy change. *Political Studies* **48**, 1006–1025.

Rieger, E. (2000). The Common Agricultural Policy. In: *Policy-Making in the European Union*, H. Wallace and W. Wallace (eds). Oxford University Press, Oxford.

Russian Federation Ministry of Natural Resources (2002). Structure of the Central Apparatus of the Russian Federation Ministry of Natural Resources, 2002.

Ruut, J. (2001). State of the Environmental Management in Estonia.

Sakwa, R. (2002). *Russian Politics and Society*. Routledge, London and New York.

Sakwa, R. and Stevens, A. (eds) (2000). *Contemporary Europe*. Macmillan Press Ltd, London.

Shaw, D. (1999). *Russia in the Modern World*. Blackwell Publishers Inc., Oxford.

Shlossberg, L. (2002). *Pskovskaya Oblast na Rubezhe Vekov: Obschestvo, Vlast, Politika 1998–2002 (The Pskov Region between Two Centuries: Society, Power, and Politics 1998–2001)*. Vozhrozhdenie, Pskov.

Stoner-Weiss, K. (2002). Soviet solutions to post-Soviet problems: has Vladimir Putin really strengthened the federal center? PONARS Policy Memo No. 283, 2002.

Turnock, D. (2001). *East Central Europe and the Former Soviet Union: Environment and Society*. Arnold, London.

Wallace, H. and Wallace, W. (eds) (2000). *Policy-Making in the European Union*. Oxford University Press, Oxford.

Wernstedt, K. (2002). Environmental Management in the Russian Federation: A Next Generation Enigma, 2002.

Yablokov, A. (2001). Environmental problems and projections in Russia. *All-Russian Special Conference for the Protection of Human Rights*, Moscow.

4

Qualitative scenarios

G.D. Gooch

4.1 INTRODUCTION

In this chapter the process by which these were constructed, and a description of the scenarios, is presented. The quantitative scenarios are described in Chapter 5. A major task in the construction of the scenarios was the definition of the number and character of the alternatives. The use of only two alternative scenarios is often considered unsuitable, as this tends to lead to an 'either or' situation. Three scenarios on the other hand can lead to a preference for the middle alternative, as the other two may be considered extreme (Schwartz, 1998). Four scenarios is a usual choice (*ibid.*), and in the Lake Peipsi case four basic in-country scenarios were constructed for each country (Russia and Estonia). It is customary to include a 'business as usual', or baseline scenario, in environmental studies, and this was also the case here. 'Business as usual' scenarios represent the future as we believe it will develop from our starting point (often the present), if future conditions are not influenced by societal policies or political interference. 'Business as usual' scenarios can be seen as a projection of present trends into the future. Alcamo (2001) claims that it is useful to construct more than one baseline scenario, but also notes that restrictions in time and resources often prevent this, and that more

 Integrated Transboundary Water Management in Theory and Practice: Experiences from the New EU Eastern Borders edited by Geoffrey D. Gooch and Per Stålnacke. ISBN: 1843390841. Published by IWA Publishing, London, UK.

than one baseline scenarios may be necessary only when working with long time spans. In the Lake Peipsi case only one qualitative baseline scenario was constructed due to time limitations and the pressing need to produce an input into the second (quantitative) stage of the scenario building process. Another form of scenario is policy scenarios or intervention scenarios, which attempt to present the results of certain known policies on future conditions. These may be used to complement baseline scenarios. In the Lake Peipsi case the policy scenarios were based on different levels of economic development and cooperation between the two countries. Three policy scenarios complemented the 'business as usual' scenario for each in-country region. Two different sets of scenarios were produced, one for each country. While these sets of scenarios were based on similar driving forces, the results of these forces were expected to differ according to the specific conditions in each individual country.

In order to construct transboundary scenarios, the country-specific scenarios were combined. However, as the number of alternatives was considered excessive (four times four country scenarios could provide 16 transboundary scenarios), only a limited number of transboundary scenarios were constructed and the final number was five. A major problem in this and many other studies of future trends is the spatial delineation of the area to be used for the scenarios. Although we were primarily interested in transboundary interaction at the regional or sub-regional level, sufficient material was not available to allow the construction of qualitative scenarios at this geographical sub-level. Instead, possible future trends at national and sub-national levels were used in the construction of the regional and sub-regional scenarios. As excessively long time scales are difficult to work with, especially at this spatial level, the time period chosen for the scenarios was 12 years, from 2003 to 2015. The choice of factors influencing the scenarios is central. In our case seven factors were identified as central. These were population, agriculture, fishing, economy, politics, administration, and environment. These factors were identified in discussions between researchers and water managers. As in the IMAGE scenarios (Alcamo and Nakicenovic 1998) the primary driving forces included population and economic growth rates. As the span of factors is relatively large, it might be claimed that even at this stage the scenarios were integrated as they included economic, social, and environmental factors. However, these qualitative scenarios were only the first step in the methodology. They were later used as an input to computer-model based quantitative scenarios (in this chapter) thus creating an integrated process combining a qualitative (social science based approach), a quantitative (natural science based approach). The approach followed the recommendations of the European Environment Agency report *'Scenarios as tools for international environmental assessments'* (Alcamo 2001) which proposes a 'story-and-simulation' (SAS) approach to developing scenarios. In this approach, basic qualitative scenarios are drafted after discussions between experts, policy makers, and stakeholders. The driving forces of these 'storylines' are then quantified, and the resulting values utilised as indicators and input for quantitative models. The results of the quantitative modelling are finally re-entered into the process and the qualitative 'storylines' revised. In an ideal situation this process is repeated as many times as necessary until the team is satisfied, after which the scenarios are distributed to stakeholders and other experts for comment. After a final revision the scenarios are presented and distributed. While a variant of the SAS model was used in the Lake Peipsi study, with the quantitative 'storyline' scenarios presented here constituting the first step of the process, the number of reformulations was restricted. The results of the quantitative modelling were however utilised as input into revised qualitative storylines, as described in Chapter 5.

4.2 SCENARIOS FOR THE RUSSIAN–ESTONIAN BORDER

When faced with the task of constructing qualitative scenarios for the Estonian–Russian border a number of factors needed to be determined. The first was the time scale of the scenarios. As noted it was decided that a 10–15-year perspective was the most practical, especially considering the high level of uncertainty that characterised conditions in the region. The next stage of the qualitative scenario building process was to identify the key variables, to ask the key questions for the future, and to determine the most probable scenarios. The seven factors, or driving forces, were combined into two key factors – economic development and transboundary cooperation – which were used to formulate the four alternative scenarios for each country. Presented as the two axes in the diagram below, these provided four alternative futures: high economic development/a high level of cooperation; high economic development/a low level of cooperation; low economic development/a high level of cooperation; and low economic development/a low level of cooperation. From these four alternative futures four basic scenarios could be developed by using the seven factors. These four in-country scenarios were named 'business as usual scenario', 'fast development scenario', 'isolation scenario', and 'crisis scenario'. These four basic scenarios were then developed for each in-country region. 'Business as usual' scenarios were constructed based on future expectations of continuing present trends. The 'fast development scenarios' for both Estonia and Russia were also determined. As far as Russia was concerned, the target domestic scenario was seen to consist of fast economic and social development. In the case of Estonia, after the decision taken by the European Union (EU) in Copenhagen December 2002 that ten new states (including Estonia) would be allowed to join the EU in May 2004 strongly influenced the creation of the 'fast development scenario' for that country, and it included fast adaptation to the EU. The 'crisis scenarios' were constructed as examples of how conditions might be expected to radically deteriorate from the present state into crisis. The 'isolationist' scenarios were for Estonia's part based on a slow, unwilling adaptation to the EU, while for Russia the 'isolationist scenario' was based on isolation from Europe and a growth of nationalist sentiment. Four scenarios were presented based on similar, but country specific, conditions in Estonia and Russia (business as usual/business as usual; fast development/fast development; crisis/crisis; isolationist/isolationist). These are presented in more detail in the following section. The resulting qualitative scenarios represented possible futures for Lake Peipsi and the Estonian–Russian border region. The four Estonian and four Russian in-country scenarios could provide up to 16 different alternative combinations. However, with so many scenarios there was a risk for 'scenario fatigue'; the capacity of many policy makers and stakeholders to consider so many alternative scenarios was considered limited. Also, some combinations appear not to fulfil the criteria that the scenarios should be at least not implausible (Stigliani *et al.* 1989). Finally, a number of alternative scenarios based on uneven in-country conditions were developed, such as the scenario that postulated that Estonia's economy would develop rapidly after membership of the EU, whilst in Russia economic, social, and political conditions would continue to deteriorate.

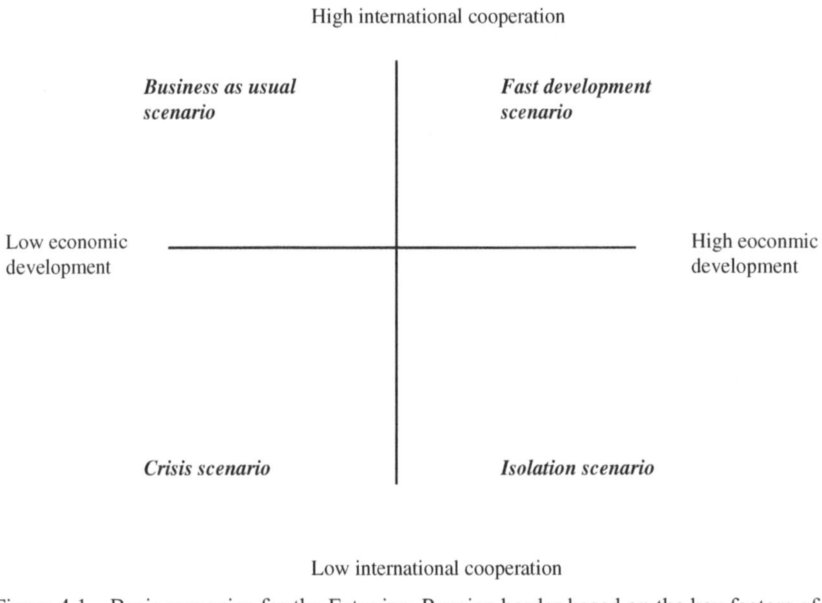

Figure 4.1 Basic scenarios for the Estonian–Russian border based on the key factors of international cooperation and economic development.

4.2.1 The 'business as usual' scenario

In the 'business as usual' scenario the population of Estonia has decreased to about 1.2 million by 2015, and the average life span has stabilised at year 2000 levels. In Russia, the average life span has stabilised at slightly below year 2000 levels and the population of the border area has decreased significantly. Agricultural production has been concentrated to a few large units in Estonia and become more industrialised. The amount of arable land has continued to decrease and is now at about 60% of the pre-1990 level. Fertiliser use has intensified somewhat over year 2000 levels and agricultural production has increased by about 25% in Estonia. Livestock levels have not risen significantly. In Russia agricultural production has remained at subsistence farming levels. The fishing industry is characterised by competition between the Estonian fishermen, who want to concentrate on larger fish such as pikeperch, and Russian fishermen who want to continue to catch small fish. The discrepancy of interests of Estonian and Russian fishermen has lead to modest non-relevant regulation of fishery allowing over-fishing in both countries. Catches have decreased and the share of non-piscivorous small fishes as well as young undersized piscivorous fishes has increased in fish populations and catches. Despite the existence of a system that limits fishing activities, the number of fishing gears used, quotas for all commercial species, minimum-landing sizes, etc. fish stocks continue to decrease as the system is not efficiently enforced. The enforcement system continues to be handicapped by corruption and a lack of sufficient control.

The economy of the region is growing at a rate of about 2–3% on the Estonian side thanks to a modest amount of foreign direct investment (FDI), domestic investment, and initiatives from the EU. A relatively small number of joint Estonian– Russian ventures have continued to develop slowly. The infrastructure continues to improve and the standard of living continues to increase slowly in Estonia, at least for a majority of the population. Tourism is well developed in the capital Tallinn and there are a small number of nature tourist enterprises in the Lake Peipsi region. On the Russian side of the border investments in production are still very low, and the economy is dependent on exploitation of natural resources. Development is not equally distributed between regions, and this has led to increased economic stratification and unequal regional development. Politically, both sides of the Russian–Estonian border have experienced a period of relative stability.

In Estonia, the country's entry into the EU in May 2004 did not result in any major changes to the country's political system. The number of political parties in Estonia has decreased but is still comparatively unstable. Voters have become bored with politics, as they feel that politics cannot make a major mark in their lives. The expectations of economic growth of the 1990s have not been fulfilled, and there is growing discontent with the right-wing economic policies that have left many people worse of. Discontent is, however, mainly expressed through cynical attitudes and the way that politicians are not well respected. In Russia, the power of the president has continued to increase as the presidential role has been gradually legitimised through changes in the constitution. This has led to an even greater dominance of Moscow over the regions. The legislative (Duma) arm of politics has lost more power. The conflict between the president and the oligarchs has become contained and an uneasy truce has been agreed. Russia has not worked especially hard to develop good relations and contacts with the EU, but has not opposed some forms of cooperation. The administrative and judicial framework necessary for encouraging investment is still not in place. Within the administration, on both sides of the border there have been only slight increases in competence and it is still difficult to employ well-educated bureaucrats with a sense of responsibility. Political decisions are difficult to implement efficiently, although the most pressing policies do tend to be implemented sooner or later, all be it in a half-hearted manner.

In Estonia, the administration is still quite centralised, and local authorities have not gained more power. In Russia, the bureaucracy is now centrally steered and dependent on the central state for the little financing that it receives. EU legislation is seen by Russia as something foreign, of little relevance. The Estonian and Russian bureaucracies have little contact with each other, although there is some contact at the top levels. Many Russian local officials have not met their Estonian counterparts for many years. In the environmental sector, in Estonia water treatment and solid waste facilities have been constructed in all major settlements, but many smaller rural areas do not have satisfactory treatment facilities. The EU's Water Framework Directive (WFD) has only been partially implemented in Estonia. In Russia, much wastewater is still emitted, untreated, or insufficiently treated. Lake Peipsi is becoming eutrophic and climatic conditions such as the increasingly warm summers affect both fish and other lake communities. The risk of summer water blooms and fish-kills has increased.

4.2.2 The 'target' or fast development scenario

In the 'target' or fast development scenario, the population of Estonia increases to about 1.7 million by 2015 and the average life span has increased by some 3 years. In Russia, the average age has increased back to pre-1990 levels and the population of the border area has also increased somewhat. Agricultural production has intensified in Estonia and become industrialised and the amount of arable land has, after a decrease during the 1990s and the beginning of the 2000s, increased again, as Estonian food products are in demand. Fertiliser use has increased 100% over 1990s levels and agricultural production has increased 50% in Estonia. Livestock levels have not increased greatly but productivity has. In Russia the farmers have specialised in livestock production and have, with the help of foreign investments, constructed food-processing plants. Infrastructural improvements have also enabled the farmers to sell their products to the St. Petersburg region, which has also grown and now constitutes a major market.

The fishing industry is also thriving, thanks to controlled catches and cooperation between fishermen on both sides of the border. In Estonia, modernised techniques have replaced the more destructive fishing methods and fish farms and fish-processing plants have been developed. On the Russian side, stricter control and modern techniques, together with successful marketing of new fish products on both the domestic and EU markets, have led to a decrease in pressure on the smaller varieties of fish such as smelt, which has been mutually beneficial for fishing on both sides of the border. The development of Russian society and the improved relations with the EU and its markets has increased the demand for more valuable fishes like pikeperch and perch. The access to western fish markets has also led to the higher appreciation of valuable large fishes by the Russian government, and this has formed the basis for an agreement between Estonia and Russia on fishery regulations aiming to protect the stocks of these fishes and to avoid over-fishing. The rapid improvement of cooperation between two countries has enabled the introduction of eels as a cash crop. This has in turn led to a significant increase of fishermen incomes while at the same time creating the need for common management of the lake fishery. There is a greater level of trust and cooperation within the international fishermen community. The value of commercial catches has increased by over 50% compared with 2002 levels, but the number of fisherman has decreased, due to considerable increases in average salaries and fishing costs.

The economy in the region is growing at a rate of over 5%, thanks to an increase of FDI, domestic investment, and initiatives from the EU. Joint Estonian–Russian ventures continue to develop fast. Investments in production have increased, and there is a steady, but controlled, development of the tourist sector. Recreational fishing is now an important industry, and contributes substantially to the region's economy. Each kilo of fish caught by recreational fishermen generates approximately ten times more value to the region than each kilo harvested in commercial fishing. Development is not equally distributed, however, and has lead to increased economic stratification and unequal regional development.

Politically, both sides of the border have experienced a period of stability which has allowed the somewhat chaotic developments of the post-1990 era to be consolidated. In Estonia, the country's entry into the EU in May 2004 provided the impetus for an even stronger Europeanisation of the country's political system. The number of political parties has decreased but voters have become disillusioned with politics, as their expectations have not been fulfilled. Discontent, however, does not take a radical form, but is instead expressed as diminishing participation in elections and a major problem to recruit politicians, especially at the local levels. Economic development and an increase in wealth have led to a slight move away from the neo-liberal economic policies of the late 1900s, as female voters have begun to vote for a more socially responsible form of politics. In Russia, the power of the president has decreased from the 2000 level at the same time as the presidential role has been legitimised through an acceptance of the constitution. The legislative (Duma) arm of politics has gained more power. The oligarchs have lost some of their importance as FDI's and domestic investment demand transparency and investments in production and infrastructure. Russia has worked hard to develop good relations and contacts with the EU, and has provided the administrative and judicial framework necessary for investment.

Within the administration there has been a major increase in competence on both sides of the border and well-educated bureaucrats with a sense of responsibility now have more power to see that political decisions are successfully implemented. In Estonia, horizontal structures and new forms of governance have developed. In Russia, the bureaucracy is now reasonably well financed, and adapts to EU legislation in order to facilitate cooperation with EU countries. The Estonian and Russian bureaucracies work together to develop their competencies and to avoid competing and counter-productive initiatives. In the environmental sector, this fast development but somewhat unbalanced economic development has lead to increased environmental and social pressures and risks. In our 12 years perspective international urban wastewater directives have been implemented in both Estonia and in Russia, and in Estonia all EU environmental standards, including the WFD have been fully implemented and targets achieved.

4.2.3 The 'crisis' scenario

In the crisis scenario, the population of Estonia has decreased rapidly to about 1 million by 2015 and the average life span has decreased by some 2 years. Conflicts between ethnic Estonians and ethnic Russians have intensified as the Estonian population has decreased most rapidly. In Russia, the average age has continued to decrease and is now over 8 years under pre-1990 levels. The population of the border area has decreased dramatically and many villages, and even towns, are now deserted. Agricultural production has regressed in Estonia to predominantly subsistence farming combined with a few large but economically unviable units. The amount of arable land has decreased with 40–50% compared with 2000 levels. Fertiliser use has decreased and agricultural production has decreased by 50% in Estonia. Unsustainable land management has decreased soil productivity. In Russia, most agricultural land is deserted and only a few elderly families attempt to manage with some kind of subsistence farming. The fishing industry has almost disappeared as over-fishing on both sides of the border has caused fish stocks to collapse. There has been no investment in the fish industry for years, and fishing villages are now deserted.

The economy in the region has deflated and now grows at only about 0.5% on the Estonian side and at under 0% on the Russian side. FDI have moved away from the regions and domestic investment has failed to move to production. Low economic growth cannot guarantee the necessary climate for investments and it generates risks for investors. Initiatives from the EU have failed due to corruption and mismanagement. The tourist sector has not developed as the lack of infrastructure, hotels, etc. is not acceptable to foreign tourists, and social problems, lack of service, and lack of facilities have created a bad reputation among foreign tourists. Economic depression in industrial and agricultural sectors has caused widespread social, political, and economic problems. Politically, both sides of the border experience a period of instability. In Estonia, the countries entry into the EU led to economic problems, an increase of inequality, and a destabilisation of the political system.

The number of political parties in Estonia has increased almost to 1992 levels and is now high. Short-term minority governments replace one another. Extreme right wing and nationalist parties have developed in both Estonia and Russia and both utilise the ethnic Russian minority question in Estonia to create political instability. Voters have become totally disillusioned with politics as they have seen their standard of living deteriorate. Discontent now takes a radical and sometimes violent form. It is a major problem to recruit politicians, especially at the local levels. In Russia, the power of the oligarchs has increased and there is an open conflict with the president. The constitution is changed or ignored at will by the president. The legislative (Duma) arm of politics has been dissolved. There is an open political conflict with Estonia and many voices demand a new 'liberation' of that country. Russian relations and contacts with the EU are conflictual. The judicial framework in Russia has deteriorated, even compared with 2000 levels. Within the administration, on both sides of the border there have been major decreases in competence and it is impossible to recruit well-educated bureaucrats with a sense of responsibility. Policies cannot be successfully implemented.

On both sides of the border, the bureaucracy is insufficiently funded, and the Estonian and Russian bureaucracies have little or no contact with each other. In the environmental sector, existing water treatment plants have fallen into disrepair, and no new plants have been built. The WFD has not been implemented. Lake Peipsi has become eutrophic as wastewater plants have stopped working. Most forest areas have been clear-cut and not replanted which has lead to an increase in soil erosion. The deterioration of the administrative and judicial sectors has created a situation characterised by a lack of state control and by overexploitation of natural resources. In Russia, toxic waste treatment plants have been built which now accept dangerous waste from other countries. On both sides of the border the remaining towns have experienced the growth of urban slums and an increase in criminality. The uneven distribution of population has generated serious problems in wastewater management, drinking water supply, and other environmental protection questions.

4.2.4 The 'isolationist' scenario

In the 'isolationist' scenario, the population of Estonia has stabilised at about 1.3 million by 2015 and the average life span has stabilised at 2000 levels. In Russia, the average life span has decreased to slightly below year 2000 levels and the population of the border area has stabilised as government initiatives have encouraged development in border regions.

Agricultural production has become concentrated to a few large units in Estonia and become more industrialised, and the amount of arable land has stabilised as Estonia aims at becoming self-sufficient in food production. Fertiliser use has increased somewhat over year 2000 levels and agricultural production has increased by about 30% in Estonia. Livestock levels have increased. In Russia agricultural production has remained at subsistence levels. The fishing industry is characterised by even greater competition between the Estonian fishermen, who want to concentrate on larger fish such as pikeperch, and Russian fishermen who want to continue to catch small fish such as smelt.

The economy of the region is still growing at a rate of about 2–3% on the Estonian side thanks to a modest amount of FDI, domestic investment, and initiatives from the EU. On the Russian side growth is only about 1%. The infrastructure has only been improved on the Estonian side of Lake Peipsi and communications between Russian and Estonian towns and cities are almost non-existent. The standard of living continues to increase slowly in Estonia, at least for a majority of the population. International tourism has not developed except in Tallinn. On the Russian side investments in production are still very low, and the economy is dependent on exploitation of natural resources.

Politically, Estonia has slowly and unwillingly adapted to the EU but there are strong forces at work that aim at an Estonia withdrawal from the Union. Relations with Russia are very limited. Voters have become very nationalistic and the level of mistrust between Russia and Estonia is high. Estonian voters are negative to the EU. In Russia, the power of the president has increased and national sentiments provide support for an isolationist policy. The role of the president has increased in importance while the legislative (Duma) arm of politics now fulfils a more or less symbolic function. The oligarchs have become established and accepted as a part of Russian politics. Russia has few relations and contacts with the EU and does not encourage cooperation with Estonia. The administrative and judicial structures are undermanned and partisan.

Within the administration, there is almost no contact between the Estonian and Russian authorities and both are suspicious of the other. Competence is low and decisions strongly influenced by the central authorities. Politicians also strongly influence implementation. In both Russia and Estonia, the bureaucracy is now centrally steered and dependent on the central state for the little financing that it receives. In Estonia EU legislation is seen as something foreign, of little relevance to the country. In the environmental sector in Estonia, water treatment and solid waste facilities have been constructed in all the major towns, but emissions into the Lake and Baltic are still high. Most rural areas do not have satisfactory water treatment facilities as it is considered too expensive. In Russia much wastewater is still emitted, untreated, or insufficiently treated. The WFD has not been implemented in Estonia. Lake Peipsi has becoming eutrophic and climatic conditions such as the increasingly warm summers, affect both fish and other lake communities. The risk of summer water blooms and fish-kills has increased. There is very little interest in the transboundary Lake.

4.2.5 Fast development in Estonia – crisis in Russia (F/C scenario)

In the F/C scenario the population of Estonia increases to 1.7 million by 2015, while the population of the Russian border regions decreases. Due to the fast growth of its economy Estonia needs to import labour, and initiates a policy of controlled, time-limited, immigration of cheap labour from Russia. Both skilled and unskilled labour is encouraged to work in Estonia, and many qualified computer workers and engineers from the St. Petersburg region seek employment in Estonia.

Agricultural workers are also in demand as the growth of Estonian agriculture, despite increases in productivity, is handicapped by an aging agricultural work force. Illegal immigration is a major problem, and urban areas are faced with increased crime rates. In fishing, the Russian sector has disappeared, and fishing in the Lake is now focussed only on large fish such as pikeperch, complemented with fish farms. The Estonian economy grows at >5% as the failure of the Russian economy has severely limited the anticipated new markets in Russia. FDI into Estonia is considerable but uneasy conditions in the neighbouring Russia disencourage some investors. Politically, relations between Estonia and Russia are wary. As a member of the EU, Estonia is bound by the new decision-making procedures first presented in June 2003. These stipulate strict border controls while Estonia feels the need for immigrant workers. Russia's isolationist politics create difficulties in Estonian–Russian relations and Russia attempts to exploit Estonia's need to comply with EU regulations for its own benefit. The Estonian administration which has rapidly developed and is now a competent bureaucratic corps has difficulty cooperating with Russian counterparts, partly because of differences in competence, partly because of centrally determined Russian isolationism. The advances made by Estonian environmental protection have not been sufficient to prevent eutrophication of Lake Peipsi, as the regions on the Russian side of the border have not been able to implement environmental improvements and legislation.

4.3 DISCUSSION OF THE CONCEPTS AND FRAMEWORKS BEHIND THE SCENARIOS

The qualitative scenarios were based on existing conditions and expectations of future developments of those conditions. The ways in which those conditions were considered to interact and the ways that they were expected to develop in the future were the result of the application of theoretically informed concepts. The validity of the scenarios is therefore dependent on the explanatory and prescriptive capabilities of the concepts underlying the projections. These concepts will therefore be made more explicit and in the final sections of this chapter the underlying presumptions for each of the seven factors are presented.

4.3.1 Population

In the case of population trends three major influences have been postulated. The first is the degree of optimism of the population as regards future developments. If the parts of a population in the reproductive age believe that the future will be better than the present and that it will provide acceptable or good living conditions for any eventual offspring, then they are more likely to be inclined to have children. Their expectations will often be influenced by perceptions of economical factors, as well as levels of security and opportunity. The second influence is the rate of development of the general standard of living for the population as a whole. The experience of most European countries has been that a higher standard of living has led to a decrease in birth levels. The reasons for this may depend on a number of factors, one of which is the growth of post-material values. This concept builds upon Maslow's pyramid-of-needs theory (Maslow 1954) and has been developed by Ronald Inglehart in a number of publications (Inglehart 1977; 1990). It postulates that increases of wealth and security during childhood and adolescence will lead to the growth of generations more concerned with satisfying a sense of personal fulfilment than with obtaining the basic necessities of life such as food and clothing (which they already have). Rising standards of living in European countries have in fact been accompanied by decreasing birth rates, although a direct connection between the two factors has not been demonstrated.

4.3.2 Agriculture

Projections of agricultural conditions have been based on developments in Russia and Estonia since the end of the 1980s, and in the case of Estonia, on the expected results of membership or non-membership of the EU. During the communist era agriculture was strongly regulated by the state, and large collective farms were established. Artificial fertilisers and pesticides, as well as fuel for farm machinery, was provided by the state, and the levels of supply determined by the authorities. Levels of fertiliser use could differ considerably between farms but in general large quantities of artificial fertilisers were spread on agricultural land. The breakdown of the command system at the end of the 1980s and beginning of the 1990s left the agricultural sector without both the guaranteed markets that had purchased their products, and a ready, cheap supply of artificial fertilisers, and pesticides. These developments led to a dramatic decrease in the use of artificial fertilisers and a sharp decline in agricultural production and the amount of land used for this production. Although it is difficult to determine precisely the percentage of agricultural land that has been taken out of production, estimates of 25–30% in Estonia are generally accepted and even larger decreases in Russia. Future developments in Estonia are dependent on EU membership and the implementation of the Common Agricultural Policy (CAP). In the candidate countries in 2003 agriculture was an important sector but not nearly as technologically developed as in the EU. The Central and Eastern European Countries (CEECs) were countries with a general higher agricultural labour force compared to the EU 15-member states. A larger share of the population was employed in and dependent on agricultural markets in the CEECs than in the union. On average 22% (a total of 9.5 million workers) of the workforce in the CEECs was employed in agriculture compared to an EU 15 average of 5% (a total of 8.2 million). In Poland, the agricultural employment rate was relatively even higher with 25% of the population employed, and the agricultural land area amounted to 58.9% of the total land area, while the agriculture share of the GDP was only 3.8% (DG for Agriculture 2000). In total, with enlargement the total EU agricultural area was expected to increase from 60 to 200 million hectares and the agricultural labour force to at least double. Three CAP scenarios (based on political expectations) were produced in January 2002 and these formed the basis of the land-use scenarios.

Figure 4.2 Grassland and wetlands in the mouth region of the Emajõgi River in Estonia. (Photo: Peeter Unt (Peipsi CTC).)

4.3.3 Fishery

Until 2000 all fishing rights in Estonia were allocated according to the historical usage principle, that is, the share of the total catch that each user was allowed remained the same from year to year. However, fish harvesting increased during the 1990s and in order to deal with this excessive cropping and to increase the profitability of the sector, a fishing rights auction system was established in 2001. According to this new legislation a certain part (ca. 10%) of the fishing rights, both according to volume quotas and equipment use rights would be allocated each year through open auctions. The remainder would be allocated on the basis of recent fishing rights use history. The first auctions were carried out in the spring of 2001. The auction system was in principle a market-driven method for fishing rights distribution. Estonian small-scale fisheries, however, were considered to be not only a part of the national economy, but also an important source of employment in remote areas. Besides this, some traditional fishing methods were considered to be a part of Estonian cultural history. The results of fishing rights auctions in the region have been analysed in research conducted by Tartu University, and have shown that state income through sale of the rights has increased considerably. The auction system has not led to a concentration of the ownership of fishing rights by large companies, a development that was expected. On the contrary, the share of the smaller owners increased. The decision was taken in 2002 to cease use of the auction system.

4.3.4 Economy

In the early 1990s, Estonia embarked on radical reforms that were more far-reaching than most other transition economies. They were able to do so partly through the exclusion of non-Estonians from active participation in the political process (by not automatically granting them citizenship rights). The economic policy that followed, which included the creation of the currency board, required significant sacrifices from a large part of the population. Despite (or perhaps because of) the hardships of the 1990s, Estonia has performed more successfully than other transitional economies starting from a similar level of development. The cultural proximity to Finland meant that newly established Estonian businesses had the chance to gain experience of foreign trade on the Finnish market. While Finnish investment in Estonia during the early 1990s was not excessive, the Finnish impact on the business environment as a whole was. Estonia also had a longer tradition of commercial banking than the other Soviet Republics. Although the early commercial bank did not succeed it did begin a tradition of a banking system (which today it is mainly in foreign hands). It is considered that the most likely outcome for the future for Estonia is a continuing economic openness and thus – most likely – economic growth despite risks for setbacks. This is however, of course dependent on global economic conditions. Many prerequisites for economic development, such as a skilled labour force, modern machinery and professional management, a stable banking system, a legal framework, and infrastructural investments arc already in place. Certain parts of Estonia do not, however, have these advantages. The Russian-speaking areas in the northeast, as well as some rural areas, have not developed.

This unequal development has been incorporated into the scenarios. There is a risk that the Estonian society will be divided, perhaps not along ethnic lines, but between the haves and the have-nots. Concerning Russia, whose main source of income remains oil and gas products and raw materials, even major increases in the prices of these commodities in the world market will probably not reach the regions bordering Estonia. Russian investment is concentrated mainly to the Moscow and St. Petersburg regions or abroad. Russia is not expected to be able to implement major changes in economic policies as it lacks political and institutional capacity.

4.3.5 Politics and administration

The expected accession of the three Baltic States to the EU would move the EU's external borders eastwards up to Russia's northwest flank. It was expected that the formation of this new border would place demands both on the enlarged EU and on Russia. Many of these problems were expected to have their roots in institutional interaction at regional and local levels, and in the implementation of national, regional, and local policies. Political and administrative relations in the EU are sometimes described as a redistribution of political power in a system known as multi-level governance, yet, as was noted in Chapter 1, within the Russian Federation an opposite process of *re-centralisation* of political power took place after Putin's rise to power in 2000. The 2003 system of political and administrative management in Russia was complicated and based on hierarchical relationships between practically all existing branches of power. The highest executive level, the lower chamber of the Russian Parliament, the State Duma, the upper chamber of the parliament, the Federation Council, were all actively involved in many forms of decision-making At the level of federal districts various Ministries were active in each of the seven federal districts. At the territorial level (federal subjects: oblasts, krays) the authorities were organised in territorial units (Budarin *et al.* 2001). At local level (cities, rayons), the Ministries and the local authorities were jointly responsible for establishing bodies (Shaw 1999, p. 145). Russian environmental management has therefore an excessively complicated institutional framework with different agencies often having overlapping functions (see Chapter 3). A climate of interagency conflict, rather than co-operative arrangements, also seemed to prevail (Committee of Environmental Protection and Natural Resources of the City of Nizhniy Novgorod 2002, p. 11). In the formulation of the qualitative scenarios the processes through which policies formulated at various administrative levels were implemented were also been taken into account. The execution of many national (and international) policies takes place predominantly at the regional and local levels. The factors influencing these processes have been termed post-decisional politics (Jordan 1997) but unfortunately our knowledge of these processes is still insufficient (Jordan 1997). The qualitative scenarios presented here are based on research that has attempted to analyse the perceptions and belief systems of the decision-makers, while also attempting to identify casual connections between ideas and policy outcomes (Bieler 2001). The decision-makers' perceptions of the situation at hand have been considered central, as:

Figure 4.3 There are only two bigger towns in the basin: Pskov in Russia, with 204,000 inhabitants and Tartu in Estonia, with 98,000 inhabitants, which means that the majority of the basin population lives in small settlements. The number of people living by the lake is comparatively small, e.g., being only 27,000 in the local municipalities bordering the lake on the Estonian side. Photograph shows the shoreline of lake Peipsi with the village of Varnja. (Photo: Peeter Unt (Peipsi CTC).)

> [O]ur understanding of the actor's images and beliefs affects the further questions that we ask about that event and the behaviour that we expect of the actor in other cases (Jervis 1976, p. 30).

The importance of cognitive aspects of the politico-administrative environment has been taken into account in the scenario building, and existing images and beliefs have been taken to be based on different understandings of previous and present conditions and events. An important result of the research undertaken while developing these scenarios was the lack of communication between Estonian and Russian officials at the regional and local levels and this lack of communication also influenced the formulation of the scenarios.

4.3.6 Environment

The possible futures of environmental conditions presented in the scenarios were based on changes in exiting levels of pollution, especially phosphor and nitrogen levels. These were however speculative at this stage and the process of qualitative–quantitative–qualitative scenario building provided more detailed information (in Chapter 5). At the time it could be seen that while the introduction of wastewater treatment in towns and settlements was an important method of reducing point-source pollution the most significant factor influencing the water quality of the

Table 4.1 Estonian scenarios 2015.

Business as usual	Fast development	Crisis	Isolationist
Population			
Decreases to 1.2 million	Increases to 1.7 million/ some immigration	Decreases rapidly to 1 million/ethnic conflicts worsen	Stabilises at 1.3 million
Agriculture			
Arable land decreases/ fertiliser use increases	Intensifies fast/arable land increases/ fertiliser use increases 100%/production increases 50%	Regresses to mostly subsistence farming and extensive meat cow production	Arable land decreases slowly
Fishing			
Focus on large fish (pikeperch)/ catches decrease	Controlled catches/ modernised techniques/ fish farms and fish plants developed	Over-fishing causes a collapse and rapid decrease in fish stock	Over-fishing causes a gradual decrease in fish stock
Economy			
Grows at >3%/ more transport/ standard of living slowly increases/ tourism slightly increases/market forces rule/FDI stabilises/steady investment	Grows at <5%/exports increase/strong FDI/ strong investment in production	Grows at >1%/ depression/FDI leave country	Grows at <3%/slow increase in FDI
Politics			
Stabilised multi-party system	EU membership/multi-party system/voters disillusioned	Extreme right- and left-wing political parties development/ conflicts with ethnic minorities (Russians)/ instability/short-term minority governments	Multi-party system/ nationalist responsibility
Administration			
Competence increases slowly/ hierarchical/very little contact with Russia	Competence develops fast/horizontal structures/governance	Breakdown of competence/no recruitment/lack of coordination/ stagnation	Steady increase of competence
Environment			
Water treatment plants only in major settlements/ WFD only partially implemented	Water treatment for 90% of wastewater/ comply with WFD/ 90% of EU Directives implemented/water pollution loads decrease 30–70%/protected areas prioritised	No new water treatment plants/fast eutrophication of lake/WFD not implemented/clear-cutting of forests/ urban deterioration – city slums	Water treatment in most settlements/ WFD not implemented

Table 4.2 Russian scenarios 2015.

Business as usual	Fast development	Crisis	Isolationist
Population			
Population decreases slowly/ average life span stabilises	Average life span increases/population increases	Average life span decreases rapidly/ population decreases rapidly	Average life span decreases/ population stabilises
Agriculture			
Stagnates and decreases/no new investment	Specialises in livestock	Regresses to mostly subsistence farming	Based on subsistence farming
Fishing			
Focus on small fish/trawl net fishing declines/ catches decrease	Controlled catches/ modernised techniques/fish plants developed	No investment/ disappears	Decreases to subsistence level
Economy			
Grows at >2%/ little new investment	Grows at <5%/ tourist development/ FDI increases/ investment in production	Depression/high inflation/grows at >0/industries close/ no new investments	Grows at >2%/ slow rate of investment
Politics			
Strong president/ weak parliament/ contained conflict with oligarchs/ re-centralisation/ slow decision-making	Legislative gains more power/ oligarchs power decreases/ presidential role legitimised/contacts with EU	Oligarchs gain more power/legislative body dissolved/ open conflict between president and oligarchs/open conflict with Estonia	Presidential power increases/ little cooperation with EU/strained relations with Estonia
Administration			
Competence stays at low level/ strongly centralised/very little contact with Estonia	Sufficiently financed/develops competence/works together with Estonian/adopts to EU	Lack resources and power	No cooperation with Estonia/strictly controlled by Moscow/negative to EU/strongly centralised
Environment			
Untreated water emitted	Water treatment plants in major settlements	Lack of state control/ wastewater plants stop working/toxic waste plants built	Little interest in border and Lake conditions

Lake was the run-off from agricultural land. Land use continuation or changes were therefore expected to play a major role in the future.

The quantitative scenarios and the way in which they were constructed are presented in the following chapter.

REFERENCES

Alcamo, J. (2001). *Scenarios as Tools for International Environmental Assessments.* European Environment Agency, Copenhagen, p. 31.

Alcamo, J. and Nakicenovic, N. (1998). Long-term greenhouse gas emission scenarios and their driving forces. *Mitigation and Adoption Strategies for Global Change* **3**(2–4), 343–381.

Anon (2002). Environmental Information System in Russia – Assessment Report. (1996, May 10).

Bieler, A. (2001). Questioning cognitivism and constructivism in IR theory: reflection on the material structure of ideas. *Politics: Surveys, Debates and Controversies in Politics* **21**(2).

Budarin, V., Sedova, A., *et al.* (2001). Statute of the RF Ministry of Natural Resources. *Water Resource Management in Russia: Lake Peipsi Basin Case.* CTC, Tartu.

Committee of Environmental Protection and Natural Resources of the City of Nizhniy Novgorod (2002). Statute of Committee of Environmental Protection and Natural Resources of the City of Nizhniy Novgorod.

Crotty, J. (2002). Economic transition and pollution control in the Russian Federation: beyond pollution intensification. *Europe–Asia Studies* **54**(2), 299–316.

DG for Agriculture, T.E.C. (2000). Poland: Agriculture and Enlargement, 2002.

Inglehart, R. (1977). *The Silent Revolution: Changing Values and Political Styles among Western Publics.* Princeton University Press, Princeton, NJ.

Inglehart, R. (1990). *Culture Shift in Advanced Industrial Society.* Princeton University Press, Princeton, NJ.

Jervis, R. (1976). *Perception and Misperception in International Politics.* Princeton University Press, Princeton, NJ.

Jordan, A.G. (1997). 'Overcoming the divide' between comparative politics and international relations approaches to the EC: what role for 'post-decisional politics'? *West European Politics* **20**(4), 43–70.

Knill, C. (2001). *The Europeanisation of National Administrations.* Cambridge University Press, Cambridge.

Maslow, A.H. (1954). *Motivation and Personality.* Harper, New York.

Peters, B.G. (1988). *Comparing Public Bureaucracies: Problems of Theory and Method.* University of Alabama Press, Tuscaloosa.

Peterson, D.J. and Bielke, E. (2001). The reorganization of Russia's environmental bureaucracy: implications and prospects. *Post-Soviet Geography and Economics* **42**(1), 65–76.

Schwartz, P. (1998). *The Art of the Long View.* John Wiley & Sons, Chichester.

Shaw, D. (1999). *Russia in the Modern World.* Blackwell Publishers Inc, Oxford.

Stigliani, W.M., Brouwer, F.M., *et al.* (1989). Future environments for Europe: some implications of alternative development paths. *The Science of the Total Environment* **80**, 1–102.

Wernstedt, K. (2002). Environmental Management in the Russian Federation: A Next Generation Enigma, 2002.

5

Quantitative scenarios and modelling

D. Mourad, M. van der Perk, T. Nõges, P. Stålnacke, M. Pihlak, E. Loigu, K. Piirimäe and B. Skakalsky

5.1 USE OF MODELS IN THE SCENARIOS FOR LAKE PEIPSI

The second step of our methodology involved the quantification of the qualitative scenarios, and the use of these quantitative results by computer-based modelling tools. It is here that one of the major challenges to integrated methodology needs to be addressed. The general indications of possible future developments needed to be translated into figures that could be utilised in the modelling process, and a number of problems are apparent. First, the qualitative scenarios are based on factors and variables, some of which are not easily quantifiable. For example, political and administrative competence cannot be easily translated into the quantitative modelling process, other than to indicate a state of inertia or activity in pollution prevention, and the level and amount of nutrients. The scale of the storylines was also

 Integrated Transboundary Water Management in Theory and Practice: Experiences from the New EU Eastern Borders edited by Geoffrey D. Gooch and Per Stålnacke. ISBN: 1843390841. Published by IWA Publishing, London, UK.

problematic. The qualitative scenarios described possible developments in areas that lacked precise geographical boundaries. The major task for the modellers was therefore to determine the scale of operationalisation, and to define the spatial borders within which the modelling process would occur.

In this chapter, we utilise these quantitative models to assess the future nutrient loadings and ecological consequences for Lake Peipsi according to the qualitative scenarios defined in Chapter 4.

5.2 RIVERINE AND SOURCE EMISSION MODELLING

The options for regional economic, social and political developments discussed in Chapter 4 are expected to affect the use and management of soil and water within the Lake Peipsi drainage basin. These changes involve factors controlling the emissions and runoff of nitrogen and phosphorus to the river network draining into the lake, and include, for example, the connection rate of settlements to sewer systems and wastewater treatment plants (WWTPs), as well as levels of fertiliser application on agricultural land. To assess changes in emissions and their effects on nutrient runoff in a quantitative manner, a straightforward long-term nutrient emission and runoff model was adopted. Since the nutrient emissions and transport rates are variable in space and time, it was a logical inference to use a geographical information system (GIS) to model the past and future changes in nutrient emissions, and simulate the resulting river loads to Lake Peipsi. The spatio-temporal modelling language of the PCRaster GIS (Wesseling *et al.* 1996) was used to develop and implement the GIS-based model.

The adopted modelling approach was originally developed by De Wit (1999; 2001), and was successfully used for simulating large-scale nutrient emissions and transport in the river basins of the Rhine, Elbe and Po rivers (De Wit 1999; 2001; De Wit and Bendoricchio 2001). This approach assumes the nutrient loads at a location on a river to be the resultant of nutrient emissions and retention processes within the upstream catchment. The nutrient emissions can be estimated based on generally available data, including agricultural and population statistics for administrative units, meteorological data, spatial data sets of land use and soil texture, and a digital elevation model (DEM). The model output consists of digital maps of estimated point and diffuse source emissions and annual riverine loads for both nitrogen (Ntot) and phosphorus (Ptot). Furthermore, the implementation in GIS permits a spatial source apportionment analysis. Time steps of 5 years are used to average out seasonal variations and short-term variations due to hydrological extremes. These were adapted to the 15-year perspective of the qualitative scenarios.

The model was implemented for the Lake Peipsi drainage basin (Mourad and van der Perk 2004) using the integrated GIS-database of this region (Langaas 2002). The period between 1985 and 1999 was chosen as a reference period: for this period, population and agricultural statistics were available to estimate nutrient emissions. Moreover, a database of river discharge and water quality data for locations along Estonian rivers was available. Observed river loads of Ntot and

Ptot were used for calibration of the model parameters before using the model for future scenario calculations. After calibration, river load estimations were in harmony with measured river loads for Estonian rivers. To feed the nutrient emission and transport model for the future scenarios (2004 to 2015–2019), the four qualitative future scenario storylines presented in Chapter 4 were converted into quantitative model inputs (see Section 5.3). The values for the various model input parameters and variables for the different scenarios were established by means of expert judgement and heuristic knowledge. Using the model setup for 1985–1999 as a starting point, nutrient emissions, losses to the surface water, and river loads into the lake were predicted under each scenario. Although the scenarios do not necessarily represent what actually is going to happen, the calculation of loads under these scenarios provides valuable information, concerning the sensitivity of nutrient loads, under conditions of changing soil and land use, and management. In this way, priorities can be defined for management in order to approach the most ideal situation regarding river nutrient loads into Lake Peipsi, which in turn, as part of the third stage of the process, could be used to formulate policy recommendations.

5.2.1 From nutrient emissions to river loads

The modelling approach for the estimation of river loads of nitrogen and phosphorous in the drainage basin of Lake Peipsi consists of three successive steps (see Figure 5.1):

(1) estimating emissions,
(2) setting up a long-term hydrological model for determining water pathways and residence times,
(3) combining emissions and hydrological pathways, the routing of nutrients through the river network including storage and decay, and calculating loads.

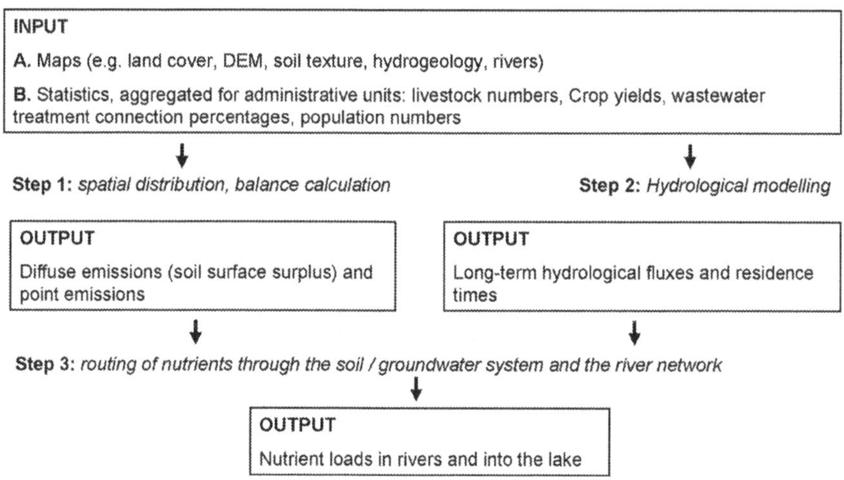

Figure 5.1 Nutrients emissions and transport modelling approach.

5.2.1.1 Step 1: Modelling diffuse- and point-sources emissions

The first step involves the estimation of nutrient emissions in the catchment and their spatial distribution based on agricultural and population statistics for each of the 5-year time steps. For the modelling of emissions, a distinction is made between point-sources emissions and diffuse-sources emissions. Nutrients from point-sources emissions are assumed to enter the surface water immediately at the emission location without any storage or decay. The main sources are here private houses, WWTPs and industries. The locations and emission rates of point sources are extracted from statistics (for major cities) or estimated by using inhabitant statistics of administration.

Nutrients from diffuse sources originate mainly from the application of manure and artificial fertilisers on agricultural land and atmospheric deposition. It is assumed that there is diffuse emission if the nutrient balance is >0. This is the case when the amount of nutrients entering the soil is greater than the amount of nutrients extracted from the soil. Inputs and outputs of nutrients in agriculture were estimated from regional agricultural statistics, such as the regional average use of artificial fertiliser, the numbers of livestock and manure excretion rates, and the harvest of the various crops and accompanying nutrient export. By combining these statistics with maps of land cover and administrative boundaries, the nutrient balance could be regionalised. Values for atmospheric deposition were borrowed from literature (Stålnacke *et al.* 2002; Loigu and Leisk 1996), and were assumed to be equally distributed over the area.

5.2.1.2 Step 2: Modelling long-term hydrological fluxes

The second step included the modelling of long-term hydrological fluxes and residence times in the hydrological pathway compartments, for each of the 5-year time steps. The hydrological sub-model was established in order to create a sound long-term water-balance runoff model of the entire catchment. It was primarily based on De Wit *et al.* (2000) and Kunkel and Wendland (2002). Taking meteorological and hydrological data as a starting point (rainfall, evapotranspiration and mean river discharges), and using relatively simple spatially available information about relief, land use and soil textures, combined with expert knowledge, the hydrological sub-model simulated the long-term distribution of water fluxes over three hydrologic pathway compartments with increasing residence time: a fast direct-surface runoff component, and two slower subsurface runoff components (intermediate and slow runoff). Average residence times were estimated from aquifer properties and drainage density.

5.2.1.3 Step 3: Transport of nutrients through the soil/ groundwater system and the river network

During the third step, results from the first two steps were combined in another submodel: diffuse emissions nutrients were routed through the soil, groundwater and river network. Nutrients originating from diffuse sources can be retained in soil and groundwater while being transported downstream to the river network, or, in the case of nitrogen, disappear by denitrification or ammonium volatilisation. These processes make the amount of nutrients entering the surface water network substantially lower than the amount that was originally emitted. Retention (both storage and decay) was assumed to be determined by local hydrology and interrelated residence times. In general, longer residence times cause a larger nutrient retention, and thus a lower nutrient input into the surface water.

In contrast to the first two steps of the modelling approach, this third step requires dynamic model calculation to account for the effect of the temporary storage of nutrients in the soil/groundwater. This implies that output from the previous time step is used as input for the next step. For each hydrological pathway pool, transport fraction factors, maximum storage capacities of the compartments, and decay factors are assigned on the basis of soil type, aquifer properties and average residence times. These decay factors account for both biological decomposition (denitrification) and retention due to geochemical processes. In this way, the flux of nutrients from all hydrological pathway compartments into the surface water is estimated. These fluxes are added to nutrient emissions from point sources to form the total input into the river network.

To account for retention and decay in the river network, transport is modelled using a local drainage direction (LDD) map, which defines a drainage direction for each grid cell. Transport fraction factors for nitrogen and phosphorous transport determine the fraction of nutrients that is transferred through the cell in downstream direction. The transport fraction factors were estimated using relatively simple assumptions. In the river network, the transport fraction factors were assumed to be inversely related to discharge and slope gradient:

$$tf = 1 - \frac{1}{1 + \left((a \cdot ((1000 \cdot S) + 1)) \cdot Q^b\right)}$$

where tf is the transport fraction (–), S is the slope gradient (–) and Q is the discharge (m^3/s). The parameters a (s/m^3) and b (–) have different values for nitrogen and phosphorus and are calibrated based on the observed river nutrient loads. The above definition of the transport fraction factor causes nutrient retention in lowland areas with small-slope gradients (long residence times) to be greater than in headwaters with large-slope gradients (short residence times).

Maps of annual river loads of total nitrogen and phosphorous (kg/year) comprise the outcome of this transport model. Values of parameters a and b are calibrated by comparing modelled and observed river nutrient loads on various locations in the drainage basin. A more detailed description of the modelling procedure is given by De Wit (1999; 2001), Mourad (2002), and Mourad and Van der Perk (2004).

5.2.2 Model implementation and input data

The basic model input for the nutrient emissions and transport model consists of a generic GIS-database of the Lake Peipsi drainage basin. This database includes digital maps of, amongst others, elevation (DEM), soil types and textures, land cover, hydrogeology and river network. In addition, map layers with the locations of measurement stations for river discharges and water quality were compiled as needed. The maps were extracted from the MANTRA-East multi-thematic GIS-database (Langaas 2002) and transformed into generic raster layers with a spatial resolution of 100 m × 100 m.

The period from 1985 until 1999, consisting of three 5-year model time steps (1985–1989, 1990–1994 and 1994–1999), was chosen as the reference period for model calibration. For this period, both model input data and model calibration data were available. Agricultural and population statistics were available at the level of a municipality, county, or entire country. In addition, an extensive dataset of river loads of total nitrogen and total phosphorus was available for locations along Estonian rivers for this period. These loads were calculated from daily discharges and periodic (monthly and bi-monthly) nutrient concentration measurements, following the guidelines of HELCOM (1994).

Most of the model parameters describing the influence of vegetation, soil and hydrogeological properties on hydrology and nutrient transport behaviour were taken from De Wit (1999; 2001), who calibrated these parameters against an extensive data set for the rivers Rhine and Elbe. The model parameters a (s/m^3) and b (–), which describe the loss in the river network, were calibrated in such a way that the simulated river nutrient loads (tons/year) were in accordance with the observed nutrient loads. For nitrogen, a amounts to 205 and b to 0.5; for phosphorous, a amounts to 109 and b to 0.9.

Although the real modelling period started in 1985, the influence on agriculture of nutrients in the various compartments had existed for quite a long time. Therefore, the first modelling time step was set to the years 1945–1949. The initial diffuse nutrient emissions were assumed to be zero for this period, and assumed to have increased linearly until 1985, for which the first detailed statistics about diffuse emission were available.

5.2.3 Model output

For the assessment of past, present and future nutrient emissions and the consequent nutrient loads into Lake Peipsi, a selection was made of results that give insight in the influence of each of the three countries in the drainage basin on nutrient loads into the lake. Moreover, results were extracted in such a way that the relative importance for nutrient loads into the lake of the two major rivers Emajõgi and Velikaya becomes clear.

The drainage basin (44,593 km^2) was divided into five regions (Figure 5.2):

- the Russian part of the Velikaya basin (22,145 km^2)
- other Russian part of the drainage basin (4203 km^2)
- the Emajõgi drainage basin, virtually completely situated in Estonia (8453 km^2)
- other Estonian part of the drainage basin (6240 km^2)
- the Latvian part of the drainage basin, mainly draining to the Velikaya (3552 km^2)

Figure 5.2 Division of the drainage basin into five parts.

5.2.4 Nutrient emissions and loads between 1985 and 1999

In general, the total gross N and P emissions in the Lake Peipsi drainage basin have been drastically reduced between 1985 and 1999 (the reasons are discussed in Chapter 2). However, there are some remarkable regional differences. The diffuse N emissions in Estonian and Latvian catchments have been reduced by more than 80% reduction between 1985 and 1999, whereas the emissions in the Russian subcatchments have been reduced by only about 50%. The N budget at the soil surface became even negative in the 1995–1999 period. The point source emissions of Ntot have only decreased substantially in the Emajõgi catchment over the three periods considered. In the other catchments, the point source emissions of Ntot have remained about the same. The diffuse P emissions in the Lake Peipsi catchment show an even greater reduction than the N emissions. In Russia, the

Table 5.1 Simulated gross emissions of total nitrogen (tons/year), 1985–1999 period.

	1985–1989	1990–1994	1995–1999
Diffuse emissions			
Russia–Velikaya catchment	206,599	150,241	108,997
Russia–other catchment	41,283	31,147	21,747
Estonia–Emajõgi catchment	83,031	32,105	13,511
Estonia–other catchment	56,117	22,574	9,920
Latvia	26,399	10,470	4,876
Total	413,429	246,537	159,051
Point emissions			
Russia–Velikaya catchment	1,640	1,570	1,496
Russia–other catchment	763	688	610
Estonia–Emajõgi catchment	1,124	949	753
Estonia–other catchment	347	349	350
Latvia	204	204	205
Total	4,078	3,760	3,414

Table 5.2 Simulated gross emissions of total phosphorus, 1985–1999 period.

	1985–1989	1990–1994	1995–1999
Diffuse emissions			
Russia–Velikaya catchment	39,858	27,616	16,711
Russia–other catchment	7,967	5,755	3,493
Estonia–Emajõgi catchment	15,302	4,220	−142
Estonia–other catchment	10,440	3,170	−341
Latvia	4,521	875	−935
Total	78,088	41,636	17,786
Point emissions			
Russia–Velikaya catchment	497	421	504
Russia–other catchment	247	174	101
Estonia–Emajõgi catchment	218	195	167
Estonia–other catchment	97	95	91
Latvia	57	56	54
Total	1,116	941	917

reduction in diffuse P emissions amounted to nearly 60%. In Estonia and Latvia, the reduction amounted to more than 100%. The point source emissions of P have only decreased in the Russian catchments other than the Velikaya catchment and in the Emajõgi catchment in Estonia (Tables 5.1 and 5.2).

5.2.5 Model calibration

After calibration of the model parameters, the simulated river nutrient loads of Estonian rivers were compared with the observed loads. Figures 5.3 and 5.4 show typically observed and simulated loads for Ntot and Ptot at two locations in the area: the Emajõgi at Tartu-Kvissental (catchment area: 7872 km^2; mean discharge: 56 m^3/s)

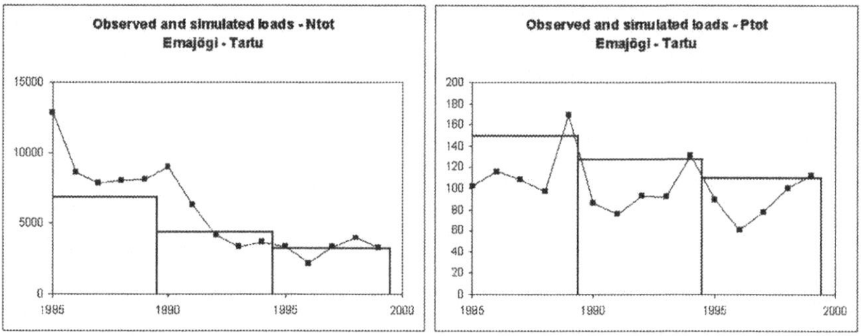

Figure 5.3 Observed (lines) and simulated (bars) Ntot and Ptot loads for the Emajõgi at Tartu.

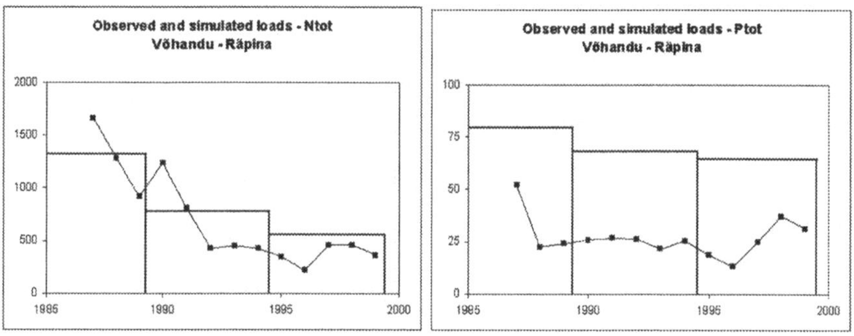

Figure 5.4 Observed (lines) and simulated (bars) Ntot and Ptot loads for the Võhandu at Räpina.

and the Võhandu river at Räpina (catchment area: 1387 km^2; mean discharge: 8 m^3/s). The model was clearly able to reproduce the massive changes in river nutrient loads during the 1985–1989 period, at least on a large (drainage basin) scale. The simulated Ntot and Ptot loads for the Emajõgi River coincide reasonably well with the observed loads, although Ptot is somewhat overestimated. Simulated loads for the Võhandu River are reasonable for Ntot, but overestimated for Ptot. Better model performance in the case of the Emajõgi catchment may be attributed to the catchment size. In the Emajõgi catchment, the variation of different load contributions averaged out to a greater extent than in the smaller Võhandu catchment, which leads to smaller uncertainties in the load estimates for the Emajõgi catchment. In general, load simulations become better with increased catchment size.

The temporal variation of observed annual Ptot loads is greater than that of Ntot. Further inspection of the original concentration and discharge data shows a strong seasonality of discharge and associated nutrient runoff. In March–April, flooding usually occurs, and discharges can increase by a factor of 3 to 6 during several weeks. Most likely, this flood period is a key period for nutrient transport, during which nutrient stored in soil and river channels are released. The large temporal variability of both discharge and nutrient concentrations results in considerable

Table 5.3 Total simulated load into Lake Peipsi (tons/year).

	1985–1989	1990–1994	1995–1999
Total Nitrogen	43,600	33,800	27,635
Total Phosphorus	1,095	890	846

uncertainties in the calculation of 'observed' nutrient loads. This effect is more pronounced for Ptot than for Ntot, because Ptot concentrations are generally low and thus more difficult to predict. Moreover, the behaviour of phosphorus in rivers is complicated because of the strong relationship between phosphorus transport and the transport of suspended sediment.

In spite of the problems noted above, the average of simulated loads coincided reasonably well with loads calculated from nutrient concentration and water discharge data, and the loads from major rivers into the lake estimated by Stålnacke *et al.* (2002).

It can be concluded that, although there is a general lack of nutrient load data on the Russian and Latvian sides of the catchment area, the model provides at least reasonable load predictions for these areas. Table 5.3 shows the simulated total inputs into Lake Peipsi during the 1985–1999 period. These values coincide reasonably with the previous estimations of Stålnacke *et al.* (2002).

5.3 IMPLEMENTATION OF THE SCENARIO STORYLINES

As the model was able to correctly simulate loads into the lake, even during the massive changes in emissions and loads that occurred during the 1980s and 1990s, it was also considered to be suitable to provide future predictions on the basis of expected input of nutrients into the drainage basin. After the completion of the modelling procedure for the 1985–1999 period, the scenario storylines used in Chapter 4 were implemented in the model.

5.3.1 Translating qualitative scenario storylines into quantitative model input

The qualitative scenarios provided in Chapter 4 were not suitable, in their original form, as input into the modelling exercises that formed the base of the quantitative scenario construction. First, a list was made of possible parameters that are important for nutrient emission and loads changes. These driving forces include *population, wastewater treatment connection rate, fertiliser use, livestock amount, crop yields, atmospheric deposition* and *amount of agricultural land.* The qualitative storylines were translated into changed values of driving forces.

In order to achieve an integrated methodology, the calibrated nutrient transport model for the period 1985–1999 (three 5-year time steps) was extended with four more time steps to cover the period 2000–2019. The scenario storylines were then translated into quantitative model input for the last time step, that is, the period 2015–2019. For this period, the modelling of nutrient emissions in the Lake Peipsi drainage basin was performed for each scenario. For the intermediate period 2000–2014, emissions were linearly interpolated in time. The hydrological submodel (Step 2) was set up using long-term average values for meteorological data. Subsequently, the accordingly calculated changing emissions served as input for the nutrient transport submodel (Step 3). The model results were reported as the average annual loads from the five regions (Figure 5.2), and the total average annual loads into the lake. Table 5.4 lists a summarizing overview of the defined changes in driving forces for the five scenarios considered.

5.4 MODELLED NUTRIENT EMISSIONS AND LOADS IN THE 2015–2019 PERIOD UNDER VARIOUS FUTURE SCENARIOS

5.4.1 Gross emission calculations

Tables 5.5 and 5.6 provide the calculated emissions that can be expected under the five scenarios. Emissions for the 1985–1999 period are also included for comparison. For both Ntot and Ptot, there are some notable outcomes: it is clear that, according to the model, total diffuse emissions will not increase again to the level of the 1980s, except for in one scenario (scenario II: Target/fast development). The lowest diffuse emissions will occur under scenario III (Crisis). Note that for all scenarios except scenarios II (Target/fast development) and V (Uneven development), negative emissions are calculated for the Estonian and Latvian parts of the drainage basin. This implies that more phosphorus will be extracted from agricultural land than added by means of manure and artificial fertilisers. For point-sources emissions, developments under various scenarios are more complicated to interpret, as on the one hand a growing economy causes emissions to increase, while on the other hand improvements in wastewater treatment efficiency, which are also expected to occur, lead to a decrease in point source emissions. In general, point source emissions are not expected to reach the 1980s maximal level again in any region. However, because of their strong connection to population levels, there is also hardly any possibility to lower these emissions significantly in the future.

5.4.2 Total riverine loads into the lake

Figures 5.5 and 5.6 show the simulated nutrient loads into Lake Peipsi for the 2015–2019 period under the five different scenarios. The loads clearly reflect the changes in emissions shown above. In general, loads will not reach 1980s values again, except for Ntot under scenario II (Target/fast development). The loads into the lake are roughly similar for all other scenarios.

Table 5.4 Changes in model driving forces under different scenarios.

Scenario	Population	WWTP connection	Fertiliser use	Livestock amounts	Crop yields	Atmospheric deposition	Amount of Agricultural land
I. Business as usual	Constant in Tartu and Pskov EST/LAT: Rural 8 % decrease RUS: 5% decrease	No changes	EST/LAT: increasing from 14 kg/ha/year N and 1.1 kg/ha P to 50 kg/ha/year N and 2 kg/ha/year P RUS: no change	EST/LAT: 10% increase RUS: no change	EST/LAT: 25% increase RUS: no change	No change	Tartu and Pskov counties: only 85% left from 1980s land Other counties: only 60% left
II. Target/fast development	Tartu and Pskov: 10% growth Rural: 5% growth	EST/LAT: only in settlements, treatment will improve one step RUS: only in settlements >10,000 inhabitants	Increasing from 14 kg/ha/year N and 1.1 kg/ha/year P to 130 kg/ha/year N and 15 kg/ha/year P	100% increase	EST/LAT: 40% growth; industrial crops: 70% increase RUS: no change	Changes from 7.7 kg/ha/year to 15 kg/ha/year (N) and from 0.05 kg/ha/year to 0.08 kg/ha/year (P)	Same amount as in 1980s
III. Crisis	Tartu and Pskov: 5% decrease EST/LAT: Rural 25% decrease RUS: 30%	EST/LAT: no change RUS: Collapse of current systems	EST/LAT: no change (14 kg/ha/year N and 1.1 kg/ha/year P) RUS: decrease with 80% to 2.8 kg/ha/year N and 0.22 kg/ha/year P	EST/LAT: 50% decrease RUS: 75% decrease, except for milk cows: 100% increase	50% decrease	No change	EST/LAT: 50% decrease RUS: 80% decrease

IV. Isolation	No change	EST/LAT: only in settlements, treatment will improve one step RUS: no change	EST/LAT: 50% increase to 21 kg/ha/year N and 1.65 kg/ha/year P RUS: no change	EST/LAT: 30% increase RUS: no change	EST/LAT: 40% increase RUS: no change	Changes from 7.7 kg/ha/year to 15 kg/ha/year (N) and from 0.05 kg/ha/yr to 0.08 kg/ha/yr (P)	EST/LAT: 10% decrease RUS: 60% decrease
V. Target/Fast Development in Estonia/Latvia, Crisis in Russia	Pskov: 5% decrease; Tartu: of 10% increase EST/LAT: Rural 5% increase RUS: Rural 30% decrease	EST/LAT: only in settlements, treatment will improve one step RUS: collapse of current systems	EST/LAT: increasing from 14 kg/ha N and 1.1 kg/ha P to 130 kg/ha N and 15 kg/ha P RUS: decrease with 80% to 2.8 kg/ha/year N and 0.22 kg/ha/year P	EST/LAT: 100% increase RUS: 75% decrease, except for milk cows: 100% increase	EST/LAT: 40% growth; industrial crops: 70% increase RUS: 50% decrease	EST/LAT: changes from 7.7 kg/ha/year to 15 kg/ha/year (N) and from 0.05 kg/ha/year to 0.08 kg/ha/year (P) RUS: no change	EST/LAT: Same amount as in 1980s RUS: 80% decrease

Table 5.5 Total nitrogen (Ntot) emissions (tons/year) for the five scenarios. Emissions for the 1985–1999 are also given for comparison.

	Simulated emissions (past)			Simulated emissions for 2015–2019 (scenarios)				
	1985–1989	1990–1994	1995–1999	I	II	III	IV	V
Diffuse emissions								
Russia – Velikaya catchment	206,599	150,241	108,997	108,108	277,388	55,329	55,329	55,329
Russia – other catchment	41,283	31,147	21,747	22,571	50,857	11,226	25,375	11,226
Estonia – Emajõgi catchment	83,031	32,105	13,511	18,522	78,410	17,108	19,542	78,410
Estonia – other catchment	56,117	22,574	9920	12,224	47,773	12,606	14,679	47,773
Latvia	26,399	10,470	4,876	6,433	26,700	6,782	6,838	26,700
Total	413,429	246,537	159,051	167,858	481,128	103,051	121,763	219,438
Point emissions								
Russia – Velikaya catchment	1,640	1,570	1,496	1,449	1,221	1,589	1,589	1,589
Russia – other catchment	763	688	610	598	489	882	610	882
Estonia – Emajõgi catchment	1,124	949	753	685	606	603	1,039	606
Estonia – other catchment	347	349	350	322	264	263	292	264
Latvia	204	204	205	188	205	153	160	205
Total	4,078	3,760	3,414	3,242	2,785	3,490	3,690	3,490

5.4.3 Source apportionment

Figures 5.7 and 5.8 again show the total Ntot and Ptot loads into the lake for the 1995–1999 period ('current situation') and for the 2015–2019 time period for each of the five scenarios, but now the load contribution from the five regions (see Figure 5.2) are shown separately. Moreover, the contribution of point and diffuse sources to the total load in the rivers has been made visible. Figures 5.7a and 5.8a

Table 5.6 Total phosphorus emissions (Ptot) emissions (tons/year) for the five scenarios. Emissions for the 1985–1999 are also given for comparison.

	Simulated emissions (past)			Simulated emissions for 2015–2019 (scenarios)				
	1985–1989	1990–1994	1995–1999	I	II	III	IV	V
Diffuse emissions								
Russia–Velikaya catchment	39,858	27,616	16,711	16,790	43,204	3,417	16,668	3,417
Russia–other catchment	7,967	5,755	3,493	3,689	8,253	824	3,671	824
Estonia–Emajõgi catchment	15,302	4,220	−1,142	−2,095	6,164	−533	−2,075	6,164
Estonia–other catchment	10,440	3,170	−341	−825	3,930	−156	−756	3,930
Latvia	4,521	875	−935	−1,347	1,095	−453	−1,431	1,095
Total	78,088	41,636	17,786	16,212	62,646	3,099	16,077	15,430
Point emissions								
Russia – Velikaya catchment	497	421	504	327	351	473	340	473
Russia – other catchment	247	174	101	97	152	256	100	256
Estonia – Emajõgi catchment	218	195	167	148	169	128	241	169
Estonia – other catchment	97	95	91	83	71	69	78	71
Latvia	57	56	54	50	54	40	54	54
Total	1,116	941	917	705	797	966	813	1,023

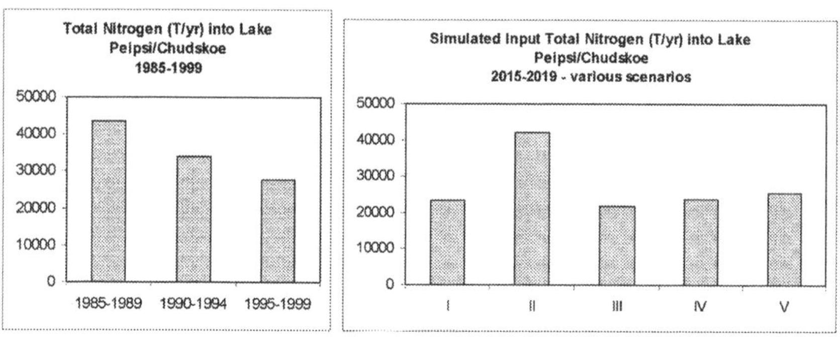

Figure 5.5 Total input of Ntot into Lake Peipsi under the five scenarios. Calibrated inputs for 1985–1999 are also given.

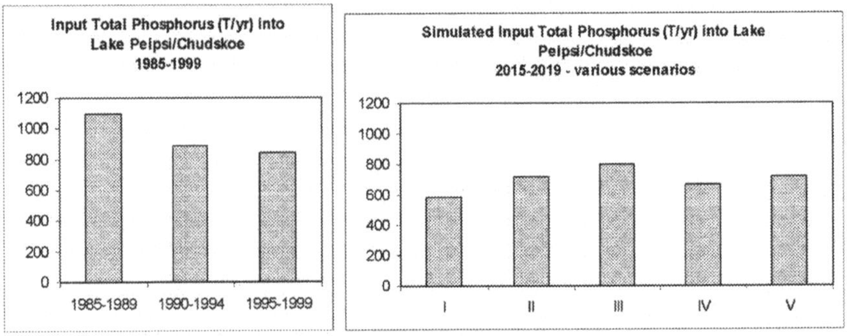

Figure 5.6 Total input of Ptot into Lake Peipsi under the five scenarios. Calibrated inputs for 1985–1999 are also given.

reveal that in the current situation, the Russian part of the catchment area, and especially the Velikaya catchment, is responsible for a vast majority of the Ntot and Ptot loads into the lake.

Less than 10% of the Ntot loads originate from point sources such as settlements and industry. This implies that agricultural activity is responsible for high loads of Ntot into the lake. For Ptot, this is not the case. Point sources are responsible for 35–65% (depending on the area) of the loads into the lake. The greatest share of point source contribution to the load (up to 65%) is found in the Velikaya River, where wastewater from the city of Pskov (200,000 inhabitants) enters the River Velikaya relatively near the southern end of the lake.

Although Figures 5.5 and 5.6 suggest that there will be no considerable changes in the total loads of Ntot and Ptot into the lake, it becomes clear from Figures 5.7 and 5.8 that significant changes will occur with respect to the relative contribution of loads from various areas and sources within the drainage basin. Scenario I (Business as usual) indicates a further reduction in both Ntot and Ptot loads, that is mainly caused by further reduction of nutrient losses from agriculture. Scenario II (Target/fast development) demonstrates an increase in Ntot losses from agriculture, while point source Ntot losses will not be significantly altered. For Ptot, the small increase in agricultural losses is compensated by a large reduction in losses, especially in the Russian Velikaya basin, caused by the expected improvement in wastewater treatment at Pskov. In scenario III (Crisis), agricultural losses of both Ntot and Ptot show a small decrease, while Ptot point sources losses increase somewhat in the Russian areas. Scenario IV (Isolationist) reveals the lowest total loads into the lake, because of reduction of both diffuse and point sources. The uneven development scenario (V) will not cause significant changes in comparison with the current situation.

5.5 ECOLOGICAL MODELLING OF LAKE PEIPSI

The methodology also included modelling of the water quality and ecological impacts on the lake itself. This was achieved through the application of AQUASIM

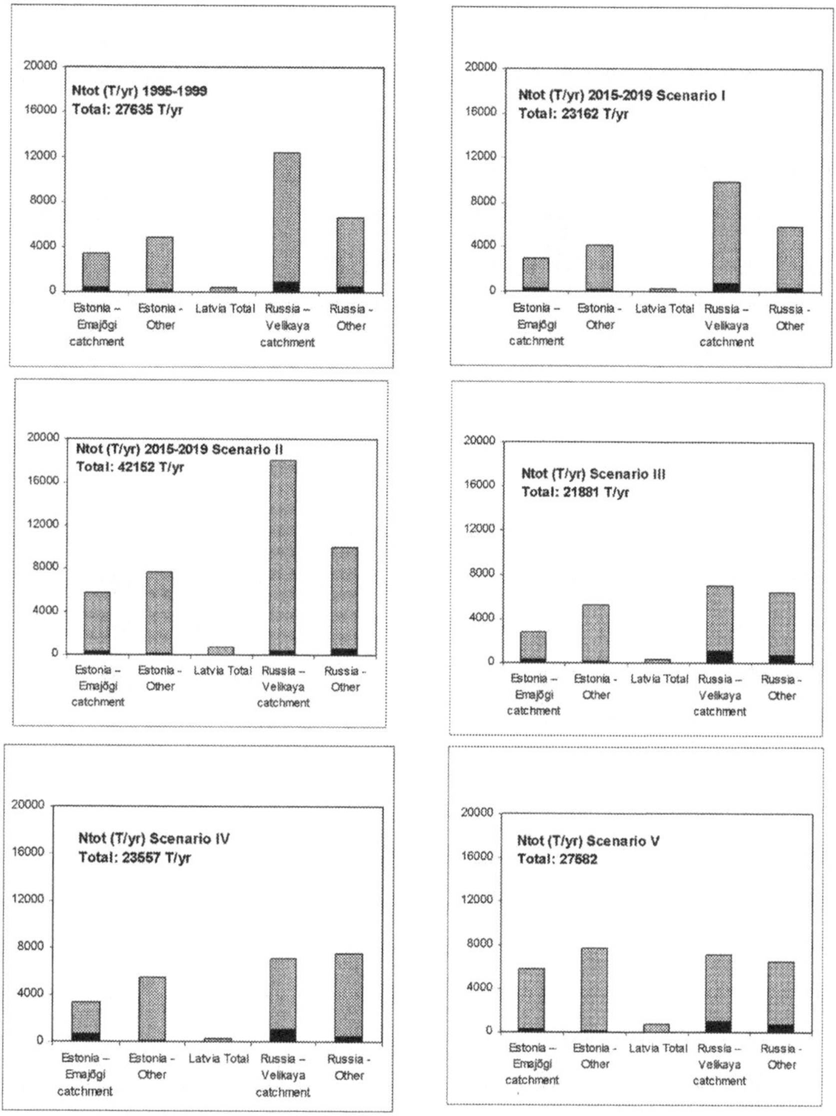

Figure 5.7 The contribution of various regions to Ntot loads into Lake Peipsi under various scenarios. Black: point sources; Grey: diffuse sources.

software (Reichert 1994). A model was composed (SHALMOD) for the ecosystem of Lake Peipsi. SHALMOD is able to simulate the growth of the main phytoplankton groups – cyanobacteria (blue-green algae) and diatoms, the dynamics of inorganic nitrogen and phosphorus and dissolved oxygen concentration. The water quality model of Lake Võrtsjärv designed by Frisk *et al.* (1999), and the values of kinetic parameters applied in the water quality model WASP-Eutro (Ambrose *et al.*

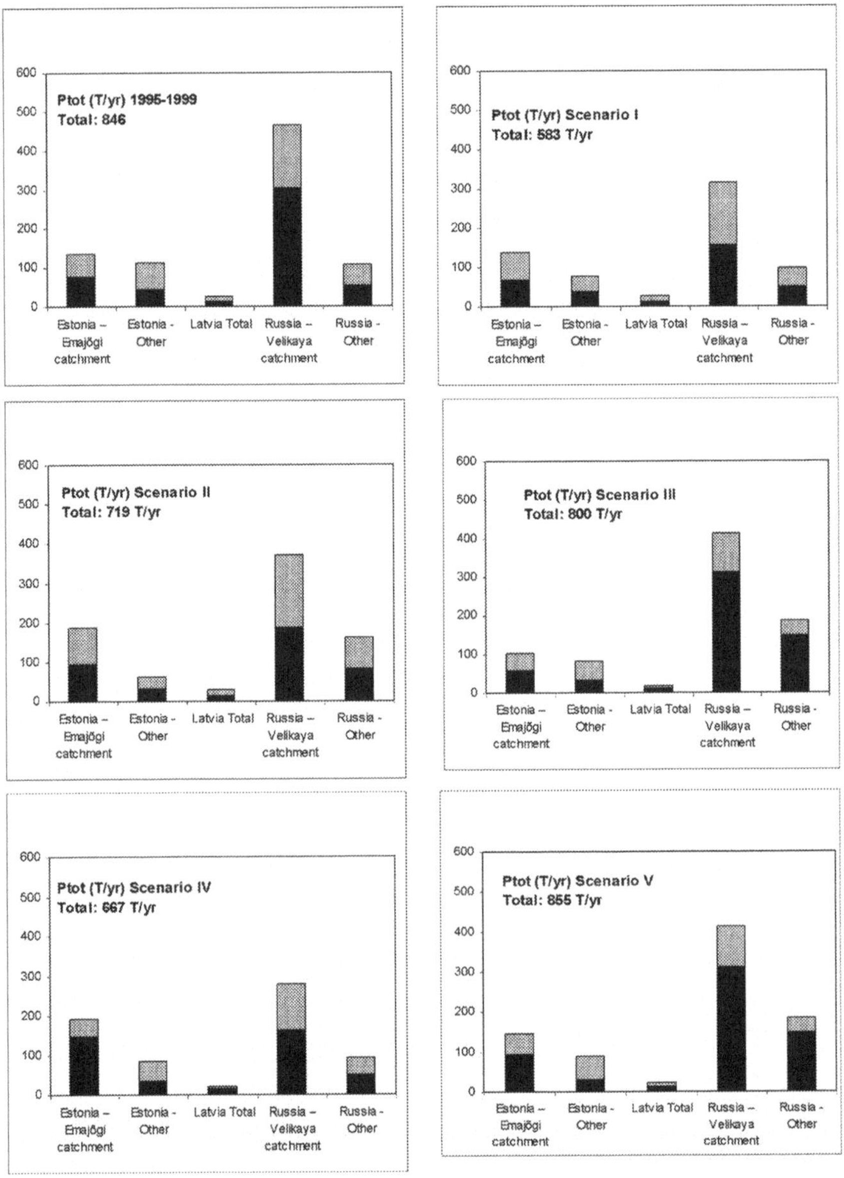

Figure 5.8 The contribution of various regions to Ntot loads into Lake Peipsi under various scenarios. Black: point sources; Grey: diffuse sources.

1993) served as the starting basis for SHALMOD. The model was calibrated for the data measured in Lake Peipsi and inflowing rivers in 1998 (Figure 5.9), thus combining the two stages of the quantitative scenario building.

The calculations of the discharges and nutrient loadings were based on the calcu-

In order to test the sensibility of SHALMOD, the dynamics of cyanobacteria and diatoms concentration in different environmental conditions were observed.

It is seen that the level of cyanobacteria tends to increase if the riverine loading of phosphorus increases and that of nitrogen decreases (Figure 5.10a, b). The response of cyanobacterial growth to the changed nitrogen loading is stronger in 1st year while the response to the changed phosphorus loading is much stronger in the 5th year after the change than in the 1st year. This may be explained through the changes of sediment phosphorus, the reduction of which at reduced phosphorus loading is also much stronger in the 5th year after the change than in the 1st year

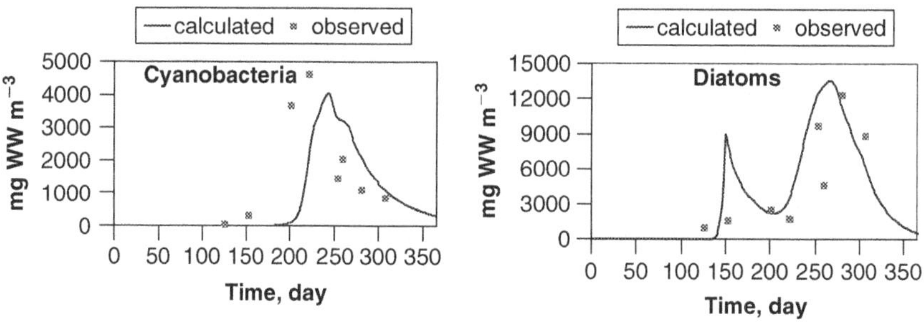

Figure 5.9 Results of calibration of cyanobacterial and diatom biomasses.

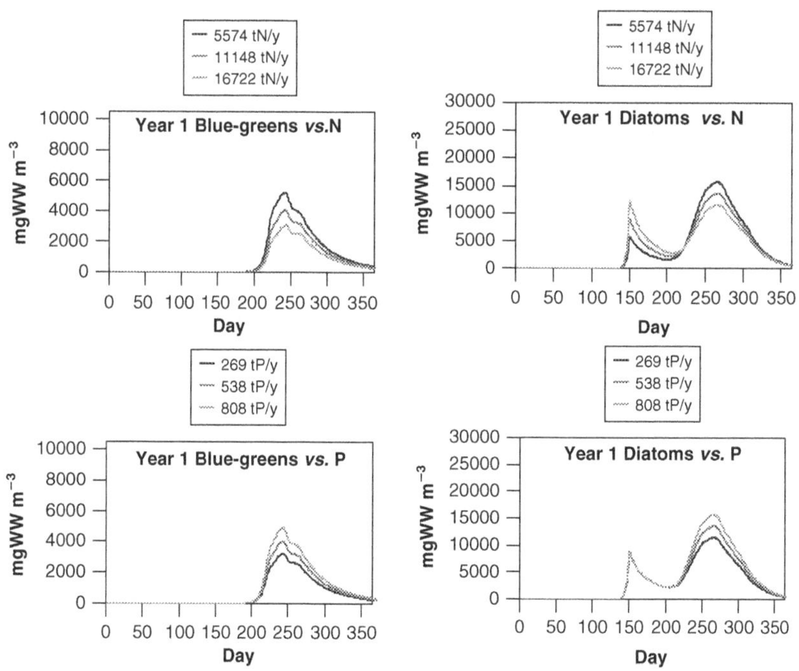

Figure 5.10a Dynamics of cyanobacteria (blue-greens) and diatoms at different N and P loading in Lake Peipsi in 1st year after change.

(Figure 5.11). It may be concluded, that it takes time for the lake to remove phosphorus, which is accumulated in the sediments.

Net production (Net), gross production (G) and annual maximum concentrations (M) of diatoms and cyanobacteria seems to be strongly dependent on the riverine loading of inorganic phosphorus, while the dependence of net production on nitrogen loading is not so strong (Figure 5.12).

Another result is the changes of water level caused by changes of riverine discharges seem to significally influence the dynamics of both cyanobacteria and diatoms, most probably by influencing the value of light limitation function (LLF). The shallower the lake, the better the average light conditions in the mixed-up water column, the light limitation is low, and better conditions exist for phytoplankton development. Water level also influences the concentration of reactive phosphorus (SRP) in sediments. The higher the water level, the less intensive is resuspension and the concentration of SRP in sediments is higher (Figure 5.13).

In order to test sensitivity to different climatic conditions, two very different years (1998 and 1999) were used. The summer of 1998 was relatively cold and cloudy while the summer of 1999 was sunny and one of the warmest in last 200 years. The differences in temperature and irradiance revealed that these conditions have a strong influence on the concentration dynamics of cyanobacteria and diatoms; in warmer water much higher concentrations of phytoplankton may occur (Figure 5.14).

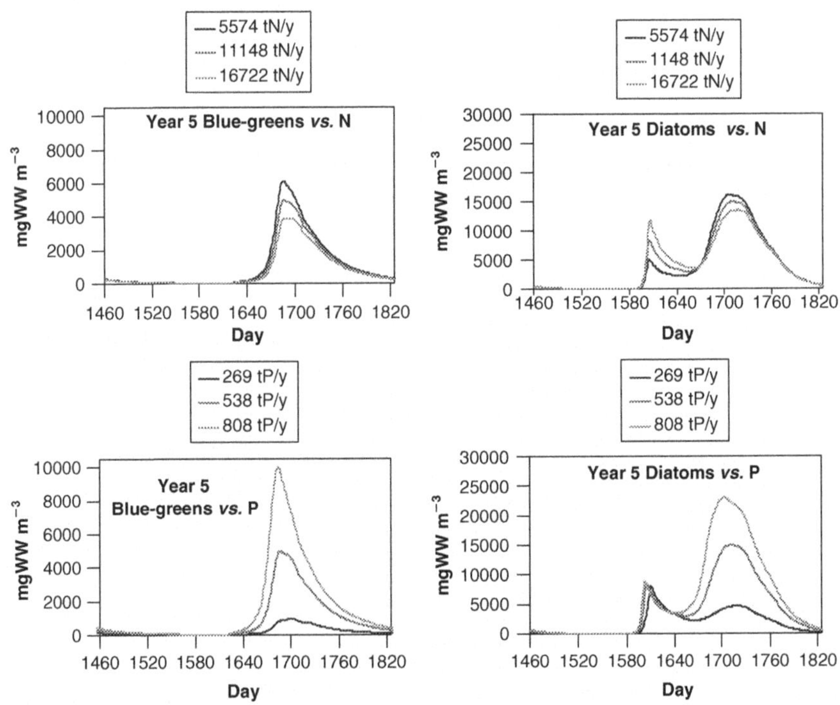

Figure 5.10b Dynamics of cyanobacteria (blue-greens) and diatoms at different N and P loading in Lake Peipsi in 5th year after change.

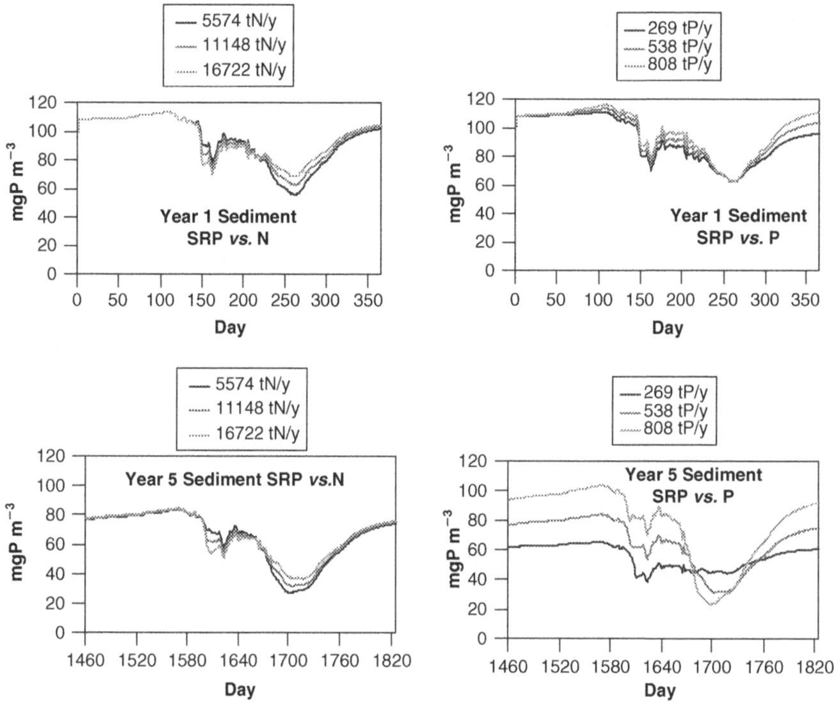

Figure 5.11 Dynamics of sediment reactive phosphorus (SRP) in Lake Peipsi at different N and P loadings in 1st and 5th year after change.

- On the basis of sensitivity analyses of SHALMOD we may conclude, that in Lake Peipsi:
- reduced riverine loading of nitrogen enchances the growth of cyanobacteria;
- increaseded riverine loading of phosphorus enchances the growth of both cyanobacteria and diatoms;
- reduction of riverine loading of phosphorus reduces the growth of both cyanobacteria and diatoms;
- phytoplankton growth is strongly influenced by weather conditions (temperature, irradiance, etc.), in warmer water much higher concentrations of phytoplankton may occur;
- hydrological conditions have significant influence on phytoplankton, higher phytoplankton concentrations occur at lower water levels.

5.5.1 Impact of different loading scenarios on the water quality of Lake Peipsi

The impact of the different nutrient loadings created by five scenarios for the socio-economic development in Estonian–Russian border region – 'Business as usual', 'Fast development', 'Crisis', 'Isolation' and 'Uneven development' (see Chapter 4 and Section 5.3) – on the water quality of Lake Peipsi was tested by SHALMOD. It is evident that the 'Crisis' scenario leads the lake closer to some kind of ecological

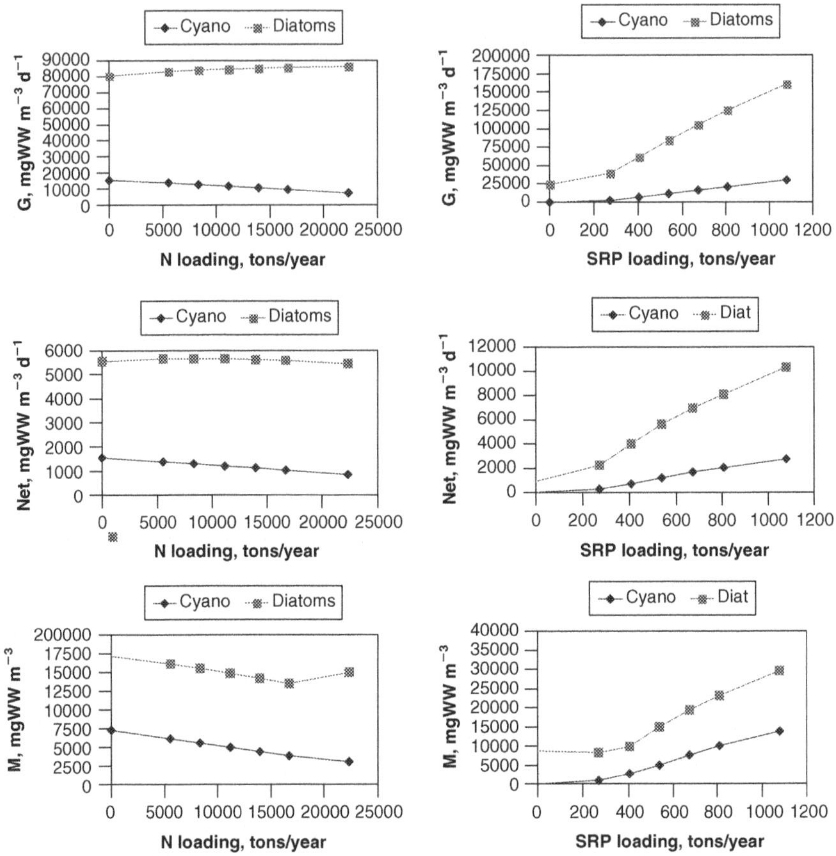

Figure 5.12 Influence of riverine nutrient loading on gross production (G), net production (Net) and annual maximum concentration (M) of cyanobacteria and diatoms in Lake Peipsi.

crisis bringing about 1.5 times higher cyanobacterial biomasses than other scenarios, especially in cold summers. In very warm summers, however, cyanobacterial biomass is higher due to climatic conditions (Figure 5.15).

5.6 DISCUSSION AND CONCLUDING REMARKS

5.6.1 Model performance

The use of GIS-based forward modelling to estimate nutrient emissions and simulate nutrient fluxes in the Lake Peipsi drainage basin appears to provide a valuable approach. As the results for the 1985–1999 reference period show, nutrient loads from large-scale (>1000 km^2) drainage basins could be simulated reasonably well. For smaller catchments, simulation is more complicated. The difference between observed and simulated loads in smaller catchments can be attributed to:

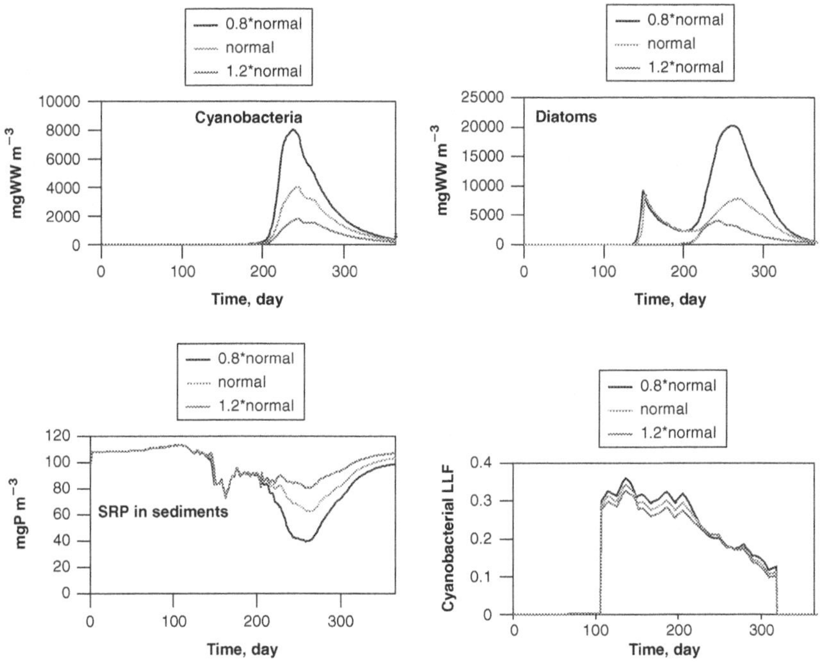

Figure 5.13 Influence of hydrological conditions (0.8*normal: input discharge 20% lower than in SHALMOD calibration; normal: input discharge as in SHALMOD calibration; 1.2*normal: input discharges 20% higher than in SHALMOD calibration) on the biomass of phytoplankton, sediment phosphorus content (SRP) and cyanobacterial LLF value.

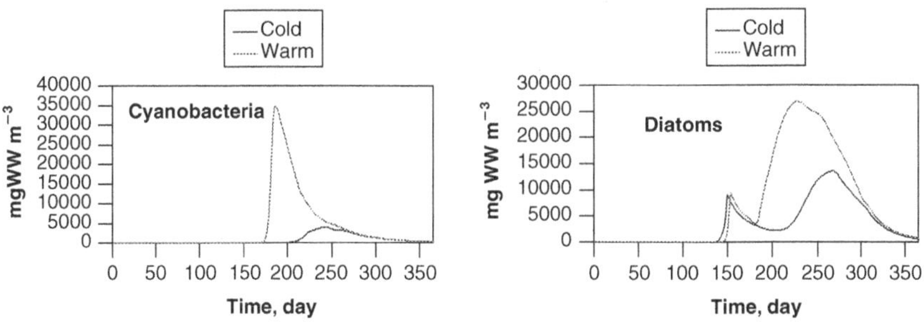

Figure 5.14 Dynamics of cyanobacteria and diatoms in cold (like 1998) and warm (like 1999) summer.

- The discrepancy between the scale of input data and the smaller scale of the catchment. Some of the input statistics are only available on the county level, while catchments are often smaller. Assumptions are made when allocating emissions to various areas that are not necessarily realistic, such as evenly distributed fertiliser application on agricultural areas. The larger the catchment, the more the effects of this variation in source areas are averaged out, and the better the prediction.

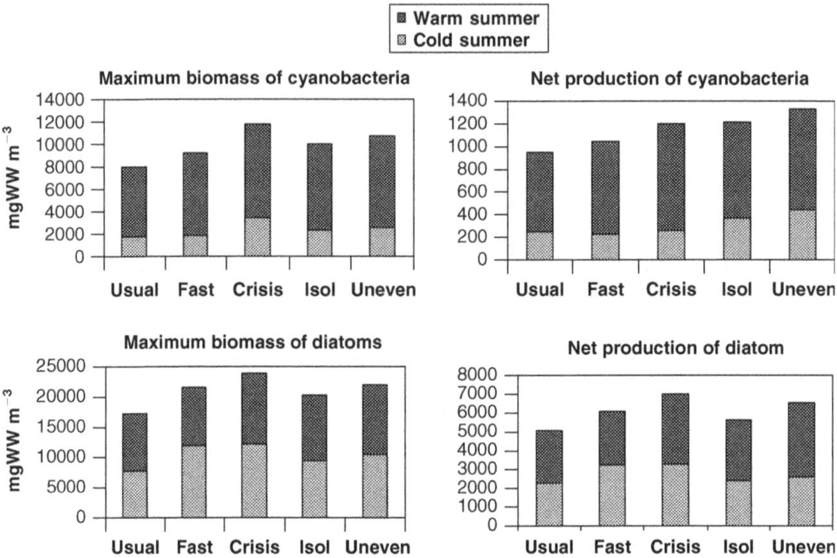

Figure 5.15 Reaction of phytoplankton to different loading scenarios in 2015–2019.

- The method of calculating 'observed' loads. For most rivers, daily discharges were available, while nutrient concentration data were sparser. These concentration data were linearly interpolated in between. Nutrient concentrations are however highly variable in time, which may cause considerable errors in the load calculations. This effect could be significant, since concentrations are known to be extremely variable in areas with pronounced spring and autumn floods under Nordic conditions.

Although the effects described above indicate that the modelling may be less suitable for small-scale predictions, our results also provide very satisfying results at the scale of the entire Lake Peipsi drainage basin. The simulated loads are in agreement with previous estimates, and the massive decrease in nutrient loads during the 1990s can be simulated quite well. One of the most important advantages of the approach is that nutrient losses can also be estimated for data-poor areas and ungauged basins. The situation in other drainage basins around the so-called 'Eastern European fringe' is comparable with the border area between Estonia/Latvia and Russia, where, for the Russian part there is often a lack of data, and even if data is available it can be very difficult to obtain (see also Chapter 2). Using conclusions and emission estimates from data-rich areas, satisfactory load simulations can be obtained for the other areas by using simple easy to obtain GIS-data layers like land use and river networks.

The forward nature of the modelling makes it very suitable for quantitative assessments of future scenario storylines. Although the simulated loads into the lake under the various scenarios that were defined for this area cannot yet be checked against observed loads, they reveal considerable information on the sensitivity of the nutrient loads to changes in emissions. Source apportionment, both spatially ('regions') and thematically (diffuse and point sources), makes it possible to set priorities for management.

5.6.2 The value of the scenario results

The simulated loads under different scenarios may look similar to each other at first sight. Indeed, under all scenarios, nutrient loads will generally decrease even more than they did during the last 15 years. Only under the Target/fast development scenario (II), will Ntot loads reach the levels of the 1980s again. Under all other scenarios, both Ntot and Ptot loads will roughly remain at the same level as now. However, a more detailed analysis reveals that significant changes can be expected in the sources of nutrients. Various regions in the drainage basin react in totally different ways with respect to nutrient emissions and loads. The main conclusions that can be drawn from the scenario calculations are:

- Ntot loads mainly originate from agriculture. Reduction of Ntot emissions from point sources will improve hygienically and water quality conditions locally, but will have a negligible effect on nutrient loads in the main rivers and therefore on trophic conditions of the lake.
- If agriculture is modernised in the future, Ntot loads will increase considerably, as the simulated loads of scenario II show. This effect will be even more pronounced if the current landscape of small-scale fields, patches of forest, low tile drainage density and rivers streaming in semi-natural valleys is altered in order to make agriculture more effective. From the perspective of nutrient losses to surface water, further amelioration of agricultural land is not advisable.
- For Ptot loads, point sources are the most important contributors. The scenario calculations show large possible variations in point source emissions and consequently Ptot loads, especially in the Russian part of the drainage basin. A considerable reduction of loads can be achieved by the improvement of wastewater treatment in the main cities on both the Estonian and Russian sides of the lake (Tartu, Pskov). Due to their relatively large size in comparison with the surrounding areas, these cities are responsible for the majority of the point-sources emissions. Moreover, their wastewater is emitted near the lake, which prevents retention in the river system.
- For the lake, reduced riverine loading of nitrogen will enhance blue-green's growth; increased riverine loading of phosphorus enhances the growth of both blue-greens and diatoms; and phytoplankton growth is strongly influenced by weather conditions: in warmer water and at lower water levels much higher concentrations of phytoplankton is expected to occur. In this case, the enhancement of blue-green blooms would be expected. More precisely, the highest N/P ratio and the less favourable conditions for cyanobacteria would be expected in case of the 'Fast development' scenario. The 'Crisis' scenario leads the lake closer to some kind of ecological crisis bringing about 1.5 times higher cyanobacterial biomasses than other scenarios, especially in cold summers. In very warm summers, however, cyanobacterial biomass is higher due to climatic conditions.
- Admitting the important role of the climatic factor as *force majeur*, the most important measures that could be done to achieve the further improvement of

water quality in Lake Peipsi would be a strong reduction of phosphorus loading from both the Estonian and Russian parts of the catchment.

REFERENCES

Ambrose, R.B., Wool, T.A., Martin, P.E. and James, L. (1993). *The water quality analysis simulation program, WASP5 Part A: Model documentation*. Environmental Research Laboratory, Athens, GA, 30613.

De Wit, M. (1999). *Nutrient fluxes in the Rhine and Elbe basins*. PhD Thesis, Faculty of Geographical Sciences, Utrecht University, Netherlands Geographical Studies 259.

De Wit, M., Meinardi, C., Wendland, F. and Kunkel, R. (2000). Modelling water fluxes for the analysis of diffuse pollution at the river basin scale. *Hydrological Processes* **14**, 1707–1723.

De Wit, M.J.M. (2001). Nutrient fluxes at the river basin scale I: the PolFlow model. *Hydrological Processes* **15**, 743–759.

De Wit, M.J.M. and Bendoricchio, G. (2001). Nutrient fluxes in the Po-basin. *The Science of the Total Environment* **273**, 147–161.

Frisk, T., Bilaletdin, Ä., Kaipainen, H., Malve, O. and Möls, M. (1999). Modelling phytoplankton dynamics of the eutrophic Lake Võrtsjärv, Estonia. *Hydrobiologia* **414**, 59–68.

HELCOM (1994). Guidelines for the third pollution load compilation (PLC-3). *Baltic Sea Environment Proceedings*, 57, 51 pp.

Kunkel, R. and Wendland, F. (2002). The GROWA98 model for water balance analysis in large river basins – the river Elbe case study. *Journal of Hydrology* **259**, 152–162.

Langaas, S. (ed.) (2002). MANTRA-East Narva river/Lake Peipsi catchment multithematic GIS database. www.mantraeast.org/gis

Loigu, E. and Leisk, Ü. (1996). Water quality of rivers in the drainage basin of Lake Peipsi. *Hydrobiologia* **338**, 25–35.

Mourad, D.S.J. (2002). *Application of GIS-based modelling to assess nutrient loads in rivers of the Estonian part of the Lake Peipsi basin*. MANTRA-East working paper. Department of Physical Geography, Utrecht University, 42 pp.

Mourad, D.S.J., van der Perk, M., Gooch, G.D., Loigu, E., Piirimäe, K. and Stålnacke, P. (2005). GIS-based quantification of future nutrient loads into Lake Peipsi/Chudskoe using qualitative regional development scenarios. *Water Science and Technology* **51**(3–4), 355–363.

Mourad, D. and van der Perk, M. (2004). Modelling nutrient fluxes from diffuse and point emissions to river loads: the transboundary Lake Peipsi/Chudskoe basin (Russia/Estonia/Latvia). *Water Science and Technology* **49**(3), 21–28.

Nõges P., Leisk Ü., Loigu E., Reihan A., Skakalski B. and Nõges T. (2003). Nutrient budget of Lake Peipsi in 1998. *Proceedings of the Estonian Academy of Sciences Biology Ecology* **4**, 407–422.

Reichert, P. (1994). *Concepts underlying a Computer Program for the Identification and Simulation of Aquatic Systems*. Swiss Federal Institute for Environmental Science and Technology (EAWAG), CH-6800, Dübendorf, Switzerland.

Stelling, G.S. (1984). *On the Construction of Computational Methods for Shallow Water Problems*. Rijkswaterstaat Communications, No. 35, Rijkswaterstaat, The Hague.

Stålnacke, P., Sults, Ü., Vasiliev, A., Skakalsky, B., Botina, A., Roll, G., Pachel, K. and Maltsman, T. (2002). Assessment of riverine loads of nutrients to Lake Peipsi, 1995–1998. *Archiv Fur Hydrobiologie* (Suppl. **141**)(3–4), 437–457.

Wesseling C.G., Karssenberg, D., Burrough, P.A. and Van Deursen, W. (1996). Integrating dynamic environmental models in GIS: the development of a Dynamic Modelling language. *Transactions in GIS* 1.1, 40–48.

WL|Delft Hydraulics, 2001: User Manual Delft3D WAQ. Delft, The Netherlands.

6

Communicating with stakeholders and the public

G. Roll and J.G. Timmerman

6.1 INTRODUCTION

Water management in a transboundary context is much more complex and multifaceted than water management within one nation state. Transboundary waters 'are interwoven with landscapes, with societies and cultures, and with political systems; international issues are always complex, always serious' – stress (Grey and Sadoff 2003). There are many dimensions of the complexity of management of transboundary waters. A sound transboundary water management system can never be an arithmetical sum of national water projects on different sides of a transboundary water basin.

Most of the issues connected with the management of waters on a transboundary level are intangible and are related to the rise of comparatively new phenomena during the last few decades. These issues include globalisation, the development of civil society and increased competition between economy sectors for limited natural resources. Due to this intangibility and also to the fact that there is only a relatively

 Integrated Transboundary Water Management in Theory and Practice: Experiences from the New EU Eastern Borders edited by Geoffrey D. Gooch and Per Stålnacke. ISBN: 1843390841. Published by IWA Publishing, London, UK.

recent history of interest in issues of water management on a transboundary scale, there is still a low awareness and understanding among water managers of the importance of addressing those intangible and complex issues as a part of the water management planning process. When the intangible complexities of water management are not properly addressed during the formation of a transboundary water regime, frustration and over-polarisation of water management issues takes place in transboundary water basins.

The major challenge of the management of transboundary waters is that the waters must be managed in the context of anarchy where there is no single government to take control. Transboundary water management is faced with the task of solving complicated problems dependent on the specific conditions created by the interaction of two or more political systems (Gooch *et al.* 2002). These specific conditions include differences or similarities in political systems of states sharing international water basins, and cultural similarities of neighbouring nations. A history of the earlier relations is also important, for example, if there are or were any political, interethnic or economic conflicts in the history of the relations of the involved states. When the political systems of the states sharing transboundary waters are close to each other and the sharing nations are also close culturally and linguistically then it is much easier to organise the cooperation. There is then more understanding and trust between those countries' political and economic elites as well as between stakeholders in these neighbouring countries. The Great Lakes cooperation between the USA and Canada or the Finnish–Norwegian cooperation on their transboundary waters are good examples of this kind of comparatively smooth cooperation of politically and culturally close nations.

However, if we argue that culturally close nations are more likely to have successful cooperation, we leave ourselves on a very weak ground. The majority of states in the world are increasingly multicultural, and besides, it is not unusual that cultural and ethnical minorities of bordering states inhabit the border areas.

Speaking of the importance of cultural connections across borders as a factor of the sustainable management of transboundary waters, it is important to keep in mind that culture is a social construct (i.e. it is something that does not exist in the objective reality). Each culture has its own 'appropriated symbols and signs which are nested in a complex set of ideas, and which is given completely through cultural assimilation, not through genetics or instinct' (Doxa 2002). Political and economic elites in bordering countries of similar cultures (i.e. sharing the same or similar sets of ideas and imperatives of behaviour) are likely to better understand each other's positions, trust each other and find joint solutions to the shared water management problems. Therefore, socially constructed images of a neighbouring state and perceptions of the nature of neighbours influences to a great extent the outcome of the cooperation in shared transboundary waters. This aspect places more focus on the importance in transboundary water basins of supporting the development of cultural links across borders, including projects that emphasise shared cultural heritages for all people living in a transboundary water region.

Culturally and politically determined perceptions of diverse issues of transboundary water management will therefore 'become realities on international rivers' (Grey and Sadoff 2003). In this context, the role of the exchange and communication of information relevant to the management of transboundary waters and the perceptions of representatives of the riparian governments become very important factors that can strongly facilitate or otherwise hinder transboundary water cooperation. The kind of information that decision-makers receive and the format of that information are very important for the outcome of the transboundary water policy-making. The main factor influencing the outcome of information communication is the source, or a sender, of information provided to policy-makers. Depending on a source of information, a decision-maker may or may not trust the information he or she receives.

At a moment of uncertainty, a decision-maker may have difficulty in identifying the form of policy that will help him or her to retain power. The way out of the dilemma is often to refer to advisors who can help to develop a solution to the problem. In this context, the existence of 'epistemic communities' or transboundary 'networks of professionals with recognised experience and competence in a particular domain and an authoritative claim to policy-relevant knowledge within that domain or issue' (Haas 1992) can be central. These are the communities or networks that the decision-makers rely on for information in the moment of uncertainty, and as such they are critically important for the success of developing transboundary water management systems. Uncertainty compels decision-makers to enlist the aid of epistemic communities. Typically, these epistemic communities will assess the problem and develop a solution to it. The epistemic community tries to act in an objective way, seeking to devise a solution that will favour the norms and beliefs held by the community. Consequently, policy-makers working within the confines of an epistemic community withdraw from the fray when the issue that is being debated does not fall within their area of specialisation. This characteristic differentiates the epistemic community from interest groups, who will stay in the fray (Ulland 2001). Transboundary epistemic communities may be formed over a long period of time by water experts and scientists working on different sides of the border on the same water-management-related issues. To stimulate the establishment of transboundary epistemic communities, more attention and support to scientific and universities' transboundary cooperation should be provided in transboundary water management. The existence of transboundary epistemic communities is an important factor to conflict prevention and the long-term sustainability of transboundary water management systems. It is therefore necessary to create greater awareness among policy-makers about the value of the transboundary scientific and educational cooperation. Better balance is needed between valuing good scientific advice to the management challenges of today on one hand, and encouraging the long-term benefits of strengthening the 'transboundary epistemic communities' on the other hand.

The next layer of complexity in the management of transboundary waters is connected with multiple uses of the water basins' natural resources and diverse interests in these basins. There are different interests of stakeholder groups that should be taken into account; these groups should be involved in the process of preparation, negotiations, and implementation of water management plans and programs from the very beginning of the process. The involvement of stakeholders in water management should be institutionalised through giving the stakeholders a possibility to participate in international water commissions, respective working groups and forums where water policy is shaped and discussed.

Managing shared water resources requires setting up institutions between riparian countries usually by means of establishing intergovernmental or interregional joint bodies that can accommodate the interests of riparian countries and other actors involved in water management in respective transboundary water basins. Joint bodies are a major instrument used in transboundary water management for resolving possible differences in interests and conflicts. The joint bodies established by transboundary agreements may vary as to format, structure or functions according to the specific circumstances in question.

As a prerequisite of stakeholder involvement in transboundary water basins, intersectoral cooperation between water managers, environmentalists, farmers, water transport businesses, industry and other interest groups should be a part of transboundary water institutions. In the countries where institutions for water management are established on a purely sectoral basis, developing intersectoral cooperation and stakeholder involvement might be a long and a painful process. Here it may be a question of changing sometimes quite strong traditional water management institutions and changing the way that water specialists think of water management to promote more integrated ways of managing transboundary waters.

The main questions that concern the institutionalisation of transboundary water cooperation are the transboundary cooperative institutions' design and their effectiveness. In the design of transboundary water institutions it is important that the institutional arrangements are tailor-made to the major environmental issues to be addressed in the region. This can be arranged through establishing temporal working groups or task forces. If a transboundary water cooperation evaluation mechanism exists, it also allows the identification of specific issues to be addressed at a given stage of development. Concerning the effectiveness of joint bodies as instruments to facilitate long-term effective transboundary water cooperation, it is worth mentioning that in some cases joint bodies have sufficient capacity to serve this role and to address conflicts and develop trust, but sometimes the capacity of joint bodies is not sufficient. It is therefore important to address factors that help or impede the effective management of transboundary waters.

The development and implementation of water management plans, including the involvement of stakeholders, should take place at three complementary levels: international, national and sub-national. At national level, national authorities develop their national policies, strategies and programs, and in doing so they have to take into account transboundary environmental risks that may affect the implementation of national policies. At the international level national agencies set up coordination mechanisms, including signing transboundary water agreements, establishing transboundary water commissions or secretariats that develop bi- or multilateral joint approaches and actions between cooperating parties on the shared waters. Public and stakeholder involvement on the national and international basin levels usually remains limited to a few large (international) non-governmental organisations (NGOs) and well-organised interest groups (Mostert 2000).

At sub-national level the participation of local governments, the private sector, NGOs, civil society institutions and various stakeholders is needed to translate these policies and programs into actions and to provide feedback. Civil society institutions active on the sub-national level are the important mechanisms for the expression of views by parties concerned with environmental issues as well as marginalised social groups who, with support, can become advocates for sustainable water use. At this level of governance it is important to involve multiple local stakeholders into larger transboundary networks of stakeholders through promoting the exchange of information and communication and helping them to develop their shared vision of the future in the transboundary water basin. It is important to provide support for the development of these networks as local stakeholders, local authorities, farmers, grassroot NGOs, etc. which usually have little knowledge and capacity to address local environmental and development problems by themselves. The potential power of these networks lies in the sharing of information and knowledge, and in the development of the capacity and skills of local organisations. On the transboundary level, support to transboundary networks is of utmost importance. Regular information exchange and communication between the stakeholders on different sides of borders creates trust; the trust and the shared vision of the future of an international basin, if it exists, is another important factor that contributes to the long-term sustainability of the management of transboundary waters.

An important issue in managing transboundary waters is financing water management. The implementation of water protection measures requires considerable financial resources, usually much higher than is available in a region. Implementation often requires the utilisation of matching finances from the respective states as well as from regional and local authorities and the private sector. It is therefore important to take into account that transboundary areas often consist of peripheral and usually less economically developed regions and that the budgets of local authorities in these areas are often poor. Usually there are not many private entrepreneurs in those border areas who are willing to put their resources into water protection either. The market for businessmen is greater in the major cities where most of the wealth of countries is usually concentrated; this is at least the case in Estonia and the other Baltic States, as well as in Russia. In this context, it is imperative to combine environmental objectives with the economic development priorities of the border regions. This is the only way to ensure the availability of financial resources in the long run for the implementation of water protection measures. Water management plans and projects should jointly address issues of economic development and environmental protection in the whole water basin and include both economic and environmental components.

6.2 MANAGING TRANSBOUNDARY WATERS ON THE EUROPEAN UNION EXTERNAL BORDERS

Along with the general transboundary water management issues that are relevant to all transboundary waters, there are specific issues of concern in transboundary water basins located on the European Union (EU) external borders. In the context of the EU economic, environmental and security policies, the EU internal borders serve multiple integration purposes including an intensification of economic networking, enhanced regional participation in policy-making, greater responsiveness to local interests in spatial planning and regional development. On the other hand, the external borders of the EU are of a different kind and are formed in an 'inter-state' context between the EU and Russia or other third states (New Independent States (NIS), Balkans, other). While practices of transboundary cooperation could be similar for both internal and external borders of the EU, the legal frameworks and institutional arrangements for the cooperation can differ considerably and this influences cooperation patterns and dynamics. For example, in areas on the EU external borders, visa-free movement of the local population across the border is often impossible while visa-free border crossing is a common practice in regions on borders between EU member states. Transboundary cooperation regimes on the EU external borders, besides their specific direct functions such as economic development (trade regimes, management of shared resources – energy, forest, oil) or environmental protection (water, biodiversity protection) also have the function of ensuring the security of the EU.

An important feature of transboundary areas on the European external borders is a growing gap in socio-economic development and living standards on the different sides of the border. If this gap grows very high it is likely that there will be little mutual understanding between the organisations and stakeholders on different sides of the border and as a result trust and cooperation across the border can decrease. The Estonian–Russian border area is one of the examples where a growing disparity in socio-economic development is taking place. This disparity will probably increase now that Estonia has become a member of the EU and Russia is on the other side of the external border of the EU. However, this gap should not be unmanageable, but in order to allow the development of understanding, trust and cooperation, it is important to provide more international funding for the Russian side of the border. This is necessary for developing infrastructure, building administrative and institutional capacity, and especially for information dissemination about the EU as well as for capacity building activities. If no practical actions are taken to manage the disparities and also to overcome them, the growing gap in socio-economic development and living standards on different sides of the border is likely to become the main bottleneck to the successful implementation of integrated water management strategies in these regions.

Transboundary water basins shared by EU countries and NIS of the former Soviet Union are usually new emerging or reconstructed borders, as it is in the case of the Estonian–Russian border. Here institutions, including procedures for communication and cooperation, are still in transition and transboundary stakeholder networks are still weak and not yet institutionalised. However, 'water management is not … a problem to be solved, but … a process that must continue in perpetuity' (Morrison 2003). To sum up, transboundary water basins should be managed interactively, that is, through regular communication and consultation among all major stakeholder groups in the basin. Neighbouring countries sharing these water basins have many differences regarding institutions, and social and economic development. In the Lake Peipsi region they are also in a process of economic and political transition. Due to these differences communicating successfully with stakeholders and the public is vital for transboundary water management. As part of the MANTRA-East project an evaluation of the use of role games as a method for improving environmental information was conducted. This study is presented in the following section.

6.3 ENVIRONMENTAL INFORMATION FOR END USERS

6.3.1 Introduction

Recent research within the MANTRA-East project and elsewhere has clearly demonstrated the significance of environmental information management in transboundary water management settings (Timmerman and Langaas 2003). With environmental information management we here mean dedicated information-related activities aimed to guide and support policy- and decision-making including operational management with an aim to resolve water quality or quantity issues of concern. Several studies indicate that environmental information may be considered the 'oil in the transboundary water management machinery' (e.g. see Enderlein 2001; Langaas and Timmerman 2003). Consistent and relevant information on the status of water systems and the associated river basin is clearly indispensable for rational and cost-effective water and river basin management. This statement has general validity for all types of river basins, but is particularly relevant for transboundary water regions. Yet, the terms 'consistent' and 'relevant' may imply different types of information for various groups of actors and stakeholders since they may operate at various levels and are concerned with a multitude of activities affecting water quality and quantity. Such activities and examples of levels may involve a consumer's choice in a grocery store on whether to buy a detergent containing phosphate or not, a farmer's decision to invest in a new technical solution for reducing the emissions of nutrient from manure, from a pig farm, or the development of a River Basin Management Plan by a transboundary water commission. Traditionally, research and development activities concerned with the use and role of environmental information in transboundary water management settings have primarily addressed the international level and in particular activities of transboundary water commissions. Timmerman and Langaas (2004) advocate that the term mind-frame could be used to describe the perspectives of various stakeholders. The information needs of the Estonian–Russian Transboundary Water Commission have been given special emphasis in the MANTRA-East project. The key argument for this is that they have assumed the task of coordinating the activities related to Water Framework Directive (WFD) implementation on the transboundary level. Still, due considerations have been given to other actors and stakeholders in the case region.

In this section we will briefly summarise some activities conducted within the case region, of which some have a research focus and others are of more technological and development nature. First, we describe an experimental study designed to examine the use and valuing of information within a transboundary water commission context and the main findings from this study. Second, we summarise some main points from a strategy developed to guide future water environmental information production, dissemination and communication activities within the Lake Peipsi region. Third, we describe briefly two interrelated activities in which the Internet is being used as a vehicle to disseminate information to various stakeholders in the region.

6.3.2 Evaluating the role and use of information in a transboundary water commission using a role game

This section provides a summary of Timmerman *et al.* (2002). Role or actor games have earlier been used to allow various water actors to play the part of another actor and thus getting insight into other perspectives and mind-frames (e.g. see Wittgren *et al.* 2000). One purpose of performing role or actor games has commonly been as an instrument to find strategies to identify remedial measures to water quality problems. In this study, we focused upon increasing the scientific understanding of the role and use of environmental information primarily developed from monitoring activities within a transboundary water commission context resembling the Estonian–Russian one concerned with the Lake Peipsi region.

The theoretical basis of the study was the recognition that information may be defined in different manners. At least six definitions of information can be identified, as listed below. The first definition describes information as related to the information user. The other definitions can be seen as part of a hierarchy in which information as useful data is the most basic and information as a constitutive force is most encompassing. In all definitions (Figure 6.1) of this hierarchy the implicit basic assumption is that information has a fixed meaning for the specific user (after Braman 1989; Rowley 1998):

(1) *Information as subjective knowledge*: In this definition, information has a variable meaning dependent upon the user's perceptions or values, and the nature of the task at hand. Information can, in this definition, be seen as potential knowledge and to convert information into knowledge, it must be integrated into an existing knowledge structure.

(2) *Information as useful data or as a 'thing'*: In this definition, data is processed for a purpose and is presented in a form that is meaningful to the recipient. Information is seen as processed and presented data. In order to communicate knowledge, information is expressed or represented in some physical way as a signal, text or communication.

(3) *Information as a resource*: In this definition, information is an objective resource, which is attainable and useable and which accordingly can be managed like other factors of production. The value of information is not readily quantifiable as its value depends on the content and use. Furthermore, information is not lost when it is given to others. This definition emphasises the uses people make of information rather than its effects upon people and society. Information as a resource links the use of information to the definition of information as a thing. This is the definition Denisov and Christoffersen (2001) use in their model where information is a resource that ultimately may cause an impact.

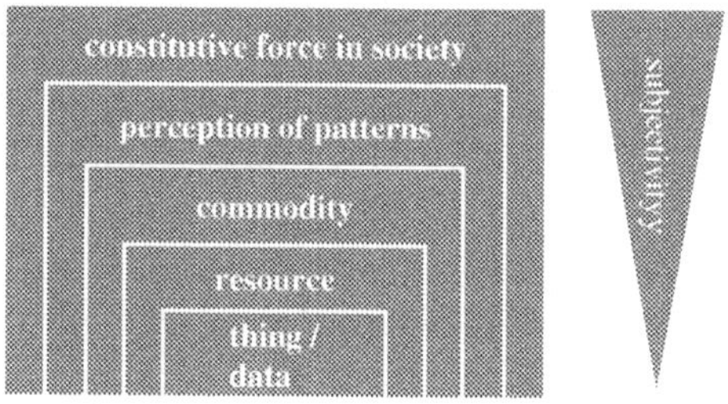

Figure 6.1 Different definitions of information.

(4) *Information as a commodity*: In this definition, the notion of information as a resource is complemented with a right to use it. This right contains a certain value that is subject to trade. The commodity notion also requires a concept of an information production chain including creation, processing, storage, transportation, distribution, etc. of information. Information gains in economic value as it passes through each stage of the chain. As compared to information as a resource, information is granted at least economic power. This is an increasingly relevant definition with the current trends towards commercialisation of public sector agencies. For example, some specific types of geographical information system (GIS data) may have very high commercial value, for example to telecommunication companies, and may tempt public sector agencies in charge of producing GIS data pricing their data high.
(5) *Information as perception of pattern*: This definition adds context to information as a commodity. Information in this perspective has a past and a future, is affected by various factors, and itself has effects. Reduction of uncertainty is one major capacity of information in this definition. Reduction of uncertainty increases the productivity of decision-making. Not only the possession of information but information itself is granted power.
(6) *Information as a constitutive force in society*: In this definition, information is a dynamic force for change in the systems within which it operates. Information is not only affected by its environment, but it is itself an actor affecting other elements in the environment. As such, information is a power in itself.

The goal of the study was to examine the type and nature of environmental information that is considered important to the policy-maker, and also to analyse the way such information is used in decision-making. The way decision-makers use information may vary depending on how they define the information (e.g. asset, commodity, infrastructure, legal obligation) and characteristics of the information provided such as:

(1) *Type of information*: What kind of information is used in what way? These can be in the form of physical, chemical, ecological, sociological, and economic information, or related to information structured according to the driving forces–pressures–stress–impact–responses (DPSIR) framework.
(2) *Presentation of information*: The way the information is presented may determine the way it is used. Presentation forms can be text, tables, maps, diagrams, etc.

(3) *Accuracy, reliability, etc.*: How much uncertainty is acceptable for the decision-maker, and how do they treat uncertainty?
(4) *Sender of information*: Does the sender of the information influence the decision? Is, for example, in a transboundary context information from the receiver's own country more valuable or trustable than information from another country?
(5) *Level of aggregation*: Will a decision-maker only use high-level, aggregated information, or will he/she be interested in detailed, local information?

This presents a wide selection of variables to the input. The way information is considered by the decision-maker was part of the observations. The information that was provided is part of the case description that has been used for the study. Given the wide range of dimensions of information that can be evaluated and considered, this study was limited to primarily consider aspects related to the type of information and the sender of the information. Furthermore, some emphasis was given to the presentation of information.

To study the use and valuing of information, a constructed and theoretical water management situation was provided to a group of selected Estonian officials that in real life also were representatives of the Lake Peipsi transboundary commission. Ideally, one would have preferred to also involve Russian officials. However, logistical and resource considerations made this impossible.

While being constructed and theoretical, the described water management situation resembled the Lake Peipsi situation quite closely so that the actors could exploit prior knowledge in the gaming situation. To determine the type of information that should be presented, it was considered that cause–effect relationships should be made clear. For this purpose, the DPSIR-framework (EEA 1998) was used. This framework describes the cause–effect chain as beginning with a *driving force* (any human or human-related activity) that causes a *pressure* (e.g. through emissions) on a water body, in turn changing the *state* of the water body. This changing state has an *impact* on the functioning of the water ecosystem and/or on the human uses of the water. This results in hopefully a societal *response* directed at any of the four elements mentioned to minimise the negative impacts.

Major driving forces were agriculture and households. Information on the type of crops, animal farms, number of households and number of cities was included. Derived from these were pressures, such as the estimated nutrient loads from crops, animal farms, and wastewater treatment plants. The status of the water was presented as concentrations of nutrients and as some trends stretching over a number of years. Impacts were described as fish kills, turbidity and oxygen concentration in the water. Possible responses (measures) were the introduction of good farming practices, connection of animal farms to wastewater treatment plants, and the improvement of treatment capacity of wastewater treatment plants. Economic and social information was also included. Economic information contained, for example the economic value of farming and fishing, costs of improving wastewater treatment capacity, etc. Social information consisted of the number of households depending on farming, fishing, etc.

The information was arranged on fact sheets. The water management situation was explained to the participants through two presentations and through the fact sheets. The task that the participants were given was to take measures in order to overcome the problems as introduced by the case. It was stated that scientists had estimated that a 40% reduction of phosphorous and nitrate loads was necessary to obtain a sustainable situation. Participants were split up into two groups: an Estonian delegation and a Russian delegation. The delegations were given time to decide upon their position. Then a 1-hour session was organised as representing the trans-boundary commission meeting in which decisions were made. The discussions were taped and transcribed.

On the basis of the transcriptions, the discussion was analysed. The points in the discussion that were considered crucial to the decision-making process were coded, as were the kind of information the decisions were based upon. The analyses of these key sections of the discussions provided insights into the policy-relevance of specific pieces of information. Special notice was given to the ways in which certain pieces of information were disputed, and the reasons that were presented for disputing them. This form of analysis provided insights into the importance that may be attached to specific forms of information and the ways that the information's quality can influence negotiations. The meeting was evaluated before closing. The participants were asked to what extent they thought this form of roleplay was realistic, and if the information provided was realistic enough, as well as if there was enough information available.

Estonia had interests in complying with the goals of the EU WFD as they would soon become an EU country. The goals as stated in the fact sheets to obtain a sustainable situation were therefore logical for the Estonian delegation. The stated necessary reduction of nutrient loads of 40% was to some extent disputed, but the overall attitude was to find solutions (measures) that would account for this reduction. The Estonian delegation used the fact sheets to learn about the driving forces and pressures, and used the information of the effectiveness of responses to come to measures. After these measures were formulated, the economic aspects (is it affordable) and to a lesser extent the social aspects were discussed. It seemed that the tendency to follow EU-regulation in combination with an improving socio-economic situation made it possible for the Estonian delegates to focus on the measures to take whereas the socio-economic factors were of secondary importance. Wastewater treatment in Estonia had reached a reasonable level and the focus in water management was shifting towards diffuse sources like agriculture. There was no reason in this situation to dispute the information presented as it was in accordance with the perceptions of the participants. The result of the discussion in the Estonian delegation was a set of measures that would lead to a considerable reduction in pollution loads. The funding of these plans was not yet established, but ideas were present to cope with this.

Russia is not aiming for EU accession and is legally not committed to the EU WFD. Furthermore, Russia has serious socio-economic problems and solutions for this situation are not in sight. Wastewater treatment in Russia is still under development and has much attention in water management. In this situation, the information as presented in the fact sheets is not in compliance with priorities as felt by the Russian delegation (made up of Estonians for the purpose of this study). One of the first issues in the discussion of the Russian delegation was consequently to dispute this information. In the fact sheets, agriculture was pointed out as the major source of pollution and was implicitly a priority issue. As the Russian delegation did not agree with agriculture being the major source, new priorities had to be found. The fact sheets were more or less disregarded, because they did not give information on the new issues. Therefore, new information had to be added to be able to discuss measures. When this was settled, information from the fact sheets was used to support the necessity of measures. Then, the socio-economic aspects of measures were discussed and finally the discussion was about summarising measures and alternatives for measures. The result of the discussion was an overview of possible measures, the effects of which were not calculated. Also measures were added that could cause an increase in pollution load. The overall effect of the proposed measures, if implemented, would never come close to the goal of 40% reduction.

The discussion in the joint meeting of the two delegations followed the results of the separate groups. The Estonian delegation could present clear measures whereas the Russian delegation had some possible measures to offer. The Russian delegation started to argue that the situation was not yet clear and that more information was needed. The Estonian delegation responded to this by asking for concrete measures. As the Russian delegation disputed the information presented, they were not able to come up with concrete proposals. Information on the state of the lake was not referred to in the discussions. Similarly, information on pressures and impacts is mostly referred to in a qualitative way. Only information on driving forces and responses is regularly used in a quantitative way to support possible measures. This implies that much of the information from physico-chemical and ecological monitoring is not used in decision-making. However, it is clear from the discussions that this information is used as a reference. For the Estonian delegation, the situation as presented in the fact sheets was recognisable and was used as a starting point for their decisions. The Russian delegation had trouble in recognising the situation as their situation and as a result started to dispute the information and asked for more information.

All in all, the participants seemed to have a sense of reality in the discussions. On the other hand, the opportunity to discuss and make decisions without any future responsibility was felt as relaxing. This implies that a real conflict would have been hard to invoke in this meeting. Based upon the evaluation of the use of the provided information (and lack thereof), the following conclusions were drawn:

- Maps are essential to describe the situation and locate specifics.
- In decision-making there is little attention for the exact figures, especially the physico-chemical and biological data. However, if such figures are not present, these will be asked for.
- If the information presented, especially information concerning the problem description, lies not within the expectation scope of the decision-maker, the decision-maker will dispute the information and ask for additional information.
- Information on the state of the water system is not directly used. Information on pressures and impacts is also little used in decision-making. In the DPSIR-framework, driving forces and responses are most used in decision-making.
- Socio-economic information, especially socio-economic implications, is indispensable for decision-making.

From these conclusions, the following recommendations were made:

- Information should, if possible and useful, be presented as geographical information in the form of maps.
- The information presented should be well founded. The information should be reliable and accurate in the perception of the decision-maker, however not necessarily very precise.
- Cause–effect relationships should be presented. These should be supplemented with information on the type and effectiveness of measures taken.
- Information for decision-making should also contain information on possible measures including their effects on the socio-economic situation of various stakeholders, that is, who will bear the cost for the various measures.
- Information for decision-making, especially the analysis of the problem, should fall within the scope of expectations of the decision-makers. For a transboundary water management situation this implies that an existing problem should be described from the viewpoints of the countries involved. Furthermore, the information should also allow for different solutions in the different countries accommodating for different implementation strategies on the basis of economic feasibility.

6.4 DEVELOPMENT AND PARTIAL IMPLEMENTATION OF AN INFORMATION AND COMMUNICATION STRATEGY FOR THE LAKE PEIPSI REGION

To guide future information-related activities in the Lake Peipsi region, an information and communication strategy (ICS) was developed. The strategy was developed by Centre for Transboundary Cooperation (CTC). CTC is an international NGO that works to promote sustainable development and cross-border cooperation in the border areas of the Baltic States and the NIS. Its main activities fall into the categories of Community Development and Civil Society and the Environment, specifically through involvement in environmental policy analysis, community and NGO development, information dissemination, communication and environmental education projects. Within the region, CTC serves as an implementing agency for a number of environmental and transboundary water management projects. CTC follows an interdisciplinary approach to the implementation of various projects that stems from a close cooperation among multiple end user organisations, including governments of the EU accession countries and EU member states, national and regional governments, municipalities and NGOs in Estonia, Latvia and Russia.

The strategy was developed both as an internal instrument for CTC's own ongoing activities in the region, and as an external support to all those projects and processes in which CTC are involved and that incorporate information-related activities aimed to resolve water-related issues, in particular, those of transboundary nature. In particular, the purpose of the ICS was to discuss and suggest mechanisms for the dissemination of expert-based environmental information concerning the

Lake Peipsi pilot study region to various stakeholder groups. This strategy is hereafter named the Lake Peipsi region ICS.

A legal and principal starting point for the development of the Lake Peipsi region ICS were the principles laid forward in the UNECE Convention on Access to Information, Public Participation in Decision-making and Access to Justice in Environmental Matters, the so-called Aarhus Convention.

Considering the rather fluid character of the term 'information', the Aarhus Convention's definition of relevant 'environmental information' was used in ICS development. According to the Aarhus Convention 'environmental information' includes information on:

- The state of the environment and its elements, such as water, air, land, soil, atmosphere, landscapes, natural sites and biological diversity, including genetically modified organisms.
- Factors such as substances, noise, radiation, and activities, including policies, legislation, plans, programs as well as data used in economic analysis influencing the state of the environment.
- The state of human health and safety, and the conditions of human life, cultural sites and buildings that might be affected by the state of environment.

Specifically, in interpreting this broad definition with regards to transboundary rivers or international lakes, public information according to the developed ICS includes:

(1) Environmental state of the transboundary waters and results of monitoring thereof, including floods and ice drifts, as well as transboundary impact.
(2) Measures taken to prevent, control and reduce transboundary impact, including water saving measures.
(3) The effectiveness of those measures.
(4) Joint and national water quality objectives.
(5) Permits issued and the conditions required to be met.
(6) Results of water effluent sampling.
(7) Results of compliance checking with water quality objectives.
(8) Results of checking water quality with permit conditions.
(9) Drafts of treaties, protocols, rules of procedure including comments from NGOs relevant for the area of application of the Water Convention or Department of Environment (DOE) specific transboundary waters.
(10) Drafts of plans and programmes, including comments by NGOs.

In an attempt to allow the public a larger role in the development of policy, the Aarhus Convention furthermore requires that the following information be included about a proposed public decision:

- The proposed activity and its possible effects on the environment, including emissions.
- A description of measures envisaged to prevent or reduce negative impacts.
- The possible decision.
- An outline of main alternatives.

- The public authority responsible for decision-making.
- Opportunities for the public to participate.
- The time and place of public hearings.
- The public authority from which necessary information can be obtained.
- The public authority to which comments, questions and proposals can be submitted.

The Convention furthermore allows that the information may be presented 'in any form, such as written, electronic, visual, aural or other material form' but it stresses that 'each country should ensure that environmental information increasingly becomes available in electronic databases easily accessible to the public', including the Internet.

A focus of the Lake Peipsi region ICS was on the relevance and importance of information dissemination, as connected to various facets of public participation.

While scholars vary in agreement over the nuances between the various forms of public participation it can be defined as 'the process by which individuals and groups come together in some way to communicate, interact, exchange information, provide input around a particular set of issues, problems, or decisions, and share in decision-making to one degree or another'. Public participation occurs at many different levels. Initially, the most basic form of participation occurs at the information dissemination level. This one-way flow of information can occur through informal seminars, information sessions, presentations, public meetings, or the translation of information into local languages or local jargon and dissemination using various media. Of course, while this form of participation requires minimal action, it is an extremely fundamental stage in more complex participatory mechanisms.

Participation can also occur in the form of consultation, which translates into two-way communication. Consultations can occur in any one of the following: participatory assessments, beneficiary assessments, consultative meetings or field visits and interviews. As the name suggests, consultations are beneficial because they garner the thoughts and perspectives of stakeholders. While information dissemination in its purest form is biased towards one perspective, consultations represent a two-way flow of ideas. The final form of public participation, and also the most intense, is empowerment. Signifying the transfer of control over decisions and resources, empowerment accords the public the highest amount of power.

Public participation with respect to the requirement of the WFD includes information supply, consultation (written and oral consultations) and active involvement where the 'active involvement' contains different collaborative participation approaches (see Figure 6.2).

As a first step in the process of developing the Lake Peipsi ICS identified the stakeholders in the region and their information needs through the following steps:

6.4.1 Step 1: Mapping the stakeholders

The main stakeholders for water use and protection in the Lake Peipsi basin were structured according to their geographical levels of operation. At the international basin level stakeholder groups include international organisations that are involved

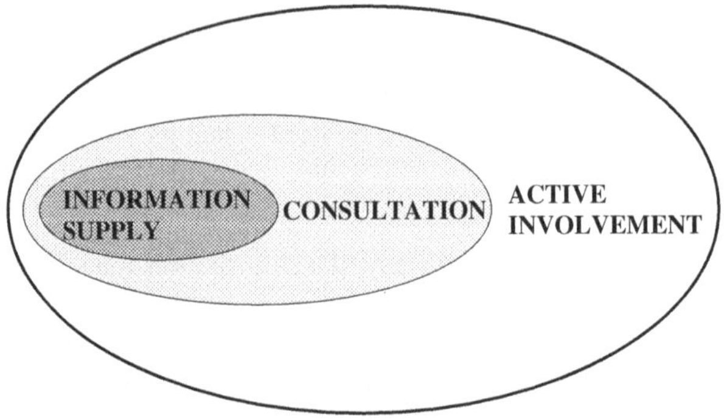

Figure 6.2 Different degrees of public participation as defined by the EU WFD Guidance Document on Public Participation. *Source*: Guidance on Public Participation (2002).

in development and implementation of policies for management of transboundary waters in Europe such as the World Bank, UNECE, UNDP, UNEP, Council of Europe, IWAC, WWF, the European Commission and other EU organisations and agencies. In the Baltic Sea region, important international organisations include the HELCOM, Council of Baltic Sea States and Baltic Sea Subregional Cooperation Council among others. On the true Lake Peipsi level there are two intergovernmental commissions: the Estonian–Russian Intergovernmental Transboundary Water Commission, and the Estonian–Russian Joint Commission for the Use of Fish Resources of Lake Peipsi, Lake Lämmi and Lake Pihkva.

There are several research groups interested in different aspects of managing waters in the Lake Peipsi basin as well. Many research projects and activities take place in the area. International development projects with a transboundary aim or scope in the region include the GEF/UNDP Lake Peipsi project, Tacis and PHARE project.

The national-level stakeholder groups include:

- national governmental institutions (sectored national agencies, such as ministries of the environment, economics, etc.);
- businesses and farmers' associations;
- national NGOs and similar associations;
- national-level universities, including researchers.

Informing and involving stakeholders on at a local level is important and can be most effective when concerning local environmental problems, such as water pollution, due to the local nature of such issues. One of the most important local-level stakeholder groups is the local government. Local governments are responsible for water management within their territory and recognise the link between development and sustainable management of water resources. Along with the local authorities, other stakeholders, such as schools, local businesses, women and youth groups, and grassroots NGOs play important roles in the implementation of water management policies.

6.4.2 Step 2: Assessing information needs of stakeholders

To assess the information needs of stakeholders several approaches were applied. The needs of the international stakeholders were quite well known through earlier activities. Thus, it was decided to place emphasis on the needs of the local-level stakeholders through a small series of dedicated information needs assessment activities. The information needs assessment activities included group interviews in several municipalities with local stakeholders, interviews with local administrators in Estonia, an anthropological study in four rural municipalities (two in Estonia and two in Russia) aimed at understanding the public's perceptions of the environment and the willingness of local stakeholders to participate in discussions and decision-making processes with regards to environmental issues on the local level. The process also included a survey of 200–300 people that explored the opinions of local people, and finally based on these public opinion surveys, eight focus group discussions in local municipalities and two citizens' juries organised on the Estonian and Russian sides of the region.

Additionally, on the combined international and national level, the before-mentioned experimental study was incorporated in the information needs assessment. Furthermore, interviews with water management experts in eight regional environmental protection departments of Estonian Ministry of the Environment were conducted.

6.4.3 Step 3: The information needs and channels

Local communities and stakeholders address local-water-related environmental problems. Therefore the perceptions of these stakeholders concerning regional development perspectives and the environmental state are important to understand. Interviews with local stakeholders showed that communication on the local level often takes place through local newspapers of which some are distributed free of charge. Local meetings were considered important to develop a dialogue between local authorities and inhabitants. A few people use Internet, mostly those involved in work of local authorities and younger people. Access to Internet is higher on Estonian side. Many people stress problems with the physical communication infrastructure. Yet, in many municipalities there are already public Internet centres or a computer class with a permanent Internet connection, or these are planned to open.

The local authorities all use the Internet on a daily basis. None, however, access specific information about the Lake Peipsi region from the Internet for the moment. All would be more interested in a comprehensive website that allowed them to do so. Topics of great importance were the economy and the environment of Lake Peipsi region. Respondents suggested that the information should be easily accessible including reports with various levels of detail, updates, news analyses and texts of international and local relevance, and that this information should be in Estonian, Russian and English language. In addition, information about culture and tourism should be included. The lack of Internet access for the majority of the general public does not necessarily stop them from receiving important information. Some are still interested in receiving information in print. In most municipalities a local newspaper is published and distributed free of charge.

Due to the poor technological infrastructure in many parts of the Lake Peipsi area, in particular the Russian part, one might conclude that Web information dissemination for the time being might not be the best information dissemination channel. However, several steps are being taken on the national and regional level to increase Internet availability in rural areas of the country. In addition, villagers have managed to find ways for information to be disseminated. Information regarding the economy, environment, agriculture, fisheries and the EU is of utmost interest.

The information needs of decision-makers are directly related to procedure, laws and regulations in question. While physico-chemical and ecological monitoring results were seldom used in decision-making, Estonian decision-makers on the national level dealt mostly with information required by the Estonian Water Act and the EU WFD. In addition, the conclusions drawn from the experimental study designed to assess information needs of national administrators on the international level (see above) were used to guide in the information needs (Timmerman *et al.* 2002):

6.4.4 Step 4: Devising the appropriate approaches

In the analysis of information needs and appropriate channels for communication, a wide variety of information needs on various levels were expressed and identified, and a wide range of optional channels for information dissemination and communication identified and discussed. Most of the information needs and communication channels are mentioned above. Two main approaches to environmental information dissemination and communication were suggested:

(1) Internet-based approaches
(2) Non-internet-based approaches

Provided a reasonable technological level and capacity, it is incontestable that the Internet is a most attractive channel for information dissemination and communication in modern societies. Although there are asymmetric development patterns and status of Internet implementation and use in Estonia and Russia, respectively, the ICS still argues that the most interesting vehicle for information dissemination is the Internet. This is in line with the viewpoints of Langaas *et al.* (2003).

It is likewise clear that alternative channels for providing pertinent information to stakeholders are required, in particular so on the local levels for those concerned with local water-related issues although often affected by and affecting the transboundary issues. Based upon the assessment activities and previous activities of CTC the following approaches were identified as useful, each with its own comparative advantages:

- *Community information sessions and campaigns*: Such interactive approaches not only work to inform the general public on environmental issues for example, but they also provide a medium for suggestions, comments and participation. On a national and regional level, especially, effective measures for disseminating information to the public and integrating stakeholders in the policy-making process include using information centres, public awareness and education programs and cultural events.

- Production and distribution of written information including brochures, information packets, pamphlets, leaflets and information booklets.
- *Citizen juries*: In these interactive 'town hall' type meetings, a randomly selected and demographically representative panel of citizens discuss the issues surrounding a particular policy area. It is of especial importance to the environmental field seeing how it is a way to collaborate both expert knowledge and public concerns, which usually is a challenge.
- *Other activities*: Peipsi CTC have coordinated various other activities and projects in the interest of furthering the public participation through the dissemination of information. These include fieldtrips, contests, conferences, workshops and tours.

In this way, the Lake Peipsi ICS provides for a consultative participation that forms the basis for future active involvement of local stakeholders. As a hopefully central instrument for the information dissemination on the lake basin environmental issues, CTC is in the process of developing a regional Lake Peipsi web portal. The purpose is to cater for the collection, processing and dissemination of information pertaining to the Lake Peipsi region. The portal, available in Estonian, Russian and English languages at http://www.peipsi.org/, has been designed to tailor-make, otherwise highly specialised environmental and sustainable development information pertaining to the Lake Peipsi Region to the needs of multiple stakeholder groups, including municipal government officials, businessmen, schools, NGOs and the local public with the aid of innovative web technologies. The home page of the portal is shown in Figure 6.3. The portal uses tools such as GIS software,

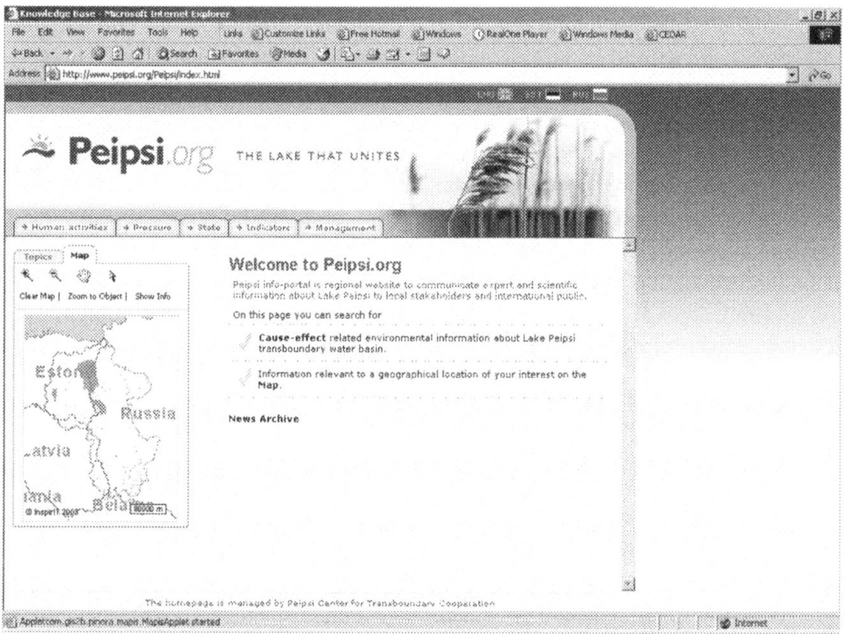

Figure 6.3 The Lake Peipsi web portal regional website aimed to communicate expert and scientific information about Lake Peipsi to local stakeholders and international public.

which allows interactive use of maps and provides important information on Lake Peipsi fauna, flora and various environmental challenges to the area. This builds upon the GIS database efforts reported in Hannerz *et al.* (2002) and Langaas *et al.* (2003). In addition, to sort through the excessive information that may exist on the web, the portal development also includes semantic region in question. Semantic Web ('next generation web') is an extension of the current web in cooperation. Information on the portal is initially structured using the DPSIR framework. There are four different ways to browse the contents of the portal: via a common hierarchical structure organised into main and sub-categories of which the main categories are accessible from every page; via a references system that is attached to every bit of information; via a direct keyword search; and via the GIS map. Visitors can run queries based on the GIS map of the Peipsi region. Altogether, the benefits of this portal will be twofold. On the one hand, it may facilitate the research process by allowing users to find the most relevant information without having to waste time for shifting through unnecessary sources. On the other hand, it may help users understand the extent of connections between different concepts and keywords encountered on a daily basis.

6.5 SUMMARY

A main water-related environmental issue of concern in the transboundary Lake Peipsi region is eutrophication. Eutrophication is a most complex environmental problem issue with a multiple set of actors and stakeholders influencing the biogeochemical cycles leading to undesirable states of nutrient concentrations in the Lake Peipsi and secondary impacts upon fish stocks. These actors and stakeholders operating from the individual household and farm level up to the international level represented by the Estonian–Russian Transboundary Water Commission and external actors such as the EU. With this in mind, and the considerable significance and role that environmental information and communication plays in transboundary river basin governance a small set of information- and communication-related activities were carried out. First, to improve the understanding of how environmental information is appreciated within the context of a transboundary river basin commission, an experimental role game study was carried out with Estonian participants. This study showed that:

- Maps are essential to describe the situation and locate specifics.
- There is little interest in precise quantification of the physico-chemical and biological status, unless they are missing or controversial.
- Information on the state of the water system is not directly used. Information on pressures and impacts is also little used in decision-making. In the DPSIR-framework, driving forces and responses are most used in decision-making.

Second, to solicit the views and opinions of other actors and stakeholders – in particular representatives of municipalities and communities – a series of other information collection activities were carried out. The aim of these was to obtain a better comprehension of information needs, and the most appropriate information and communication channels. Based upon this an ICS was developed. In the analysis of information needs and appropriate channels for communication, a wide variety of information needs on various levels were expressed and identified, and a wide range of optional channels for information dissemination and communication was identified and discussed. Two main approaches to environmental information dissemination and communication were suggested, Internet-based approaches and Non-internet-based approaches.

Third, in attempt to partially implement some components suggested in the ICS, a web portal has been designed and is now under development and continuous updation.

REFERENCES

Braman, S. (1989). Defining information: an approach for policymakers. *Telecommunications Policy* **13**(3), 233–242.

Denisov, N. and Christoffersen, L. (2001). *Impact of environmental information on decision making processes and the environment*. Occasional Paper 01, Arendal, Norway. http://www.grida.no/impact/ (accessed online at October 14 2003).

Doxa (2002). *Science and religious belief*. http://www.geocities.com/metacrock2000/science/science2_kuhn.htm

EEA (1998). *Europe's Environment: The Second Assessment*. Elsevier Science Ltd., Oxford, UK, 295 pp.

Enderlein, R.E. (2001). Bilateral and multilateral cooperation under the Water Convention. Paper presented at Workshop *Environmental Status Assessment of 10 Transboundary Rivers in Europe*, Prague, April 9–10, 2001.

Haas, P.M. (1992). Introduction: Epistemic Communities and International Policy Coordination. *International Organization* **46**, 1–37.

Hannerz, F., Langaas, S., Nilsson, S. and Tian, Y. (2002). *A multi-thematic GIS database of the transboundary Narva River/Lake Peipsi catchment in support of strategic science and management applications*. Report No. 8a, MANTRA-East Project, KTH – Royal Institute of Technology, Sweden, 50 pp.

Gooch, G., Höglund, P., Roll, G. and Aliakseyeva, N. (2002). *Review of existing structures, models and practices for transboundary water management*. MANTRA East Report D1b.

Grey, D. and Sadoff, C. (2003). Beyond the river: the benefits of cooperation on international rivers. *Balancing Competing Water Uses – Present Status and New Prospects Water Science and Technology* **47**(6), 91–96.

Langaas, S. and Timmerman, J.G. (2003). Introduction (Chapter 1). In: *Environmental Information in European Transboundary Water Management*, J.G. Timmerman and S. Langaas (eds). IWA Publishing, London.

Langaas, S., Ahlenius, H., Hannerz, F. and Nilsson, S. (2003). Towards GIS- and Internet-based information systems for transboundary river basins (Chapter 10). In: *Environmental Information in European Transboundary Water Management*, J.G. Timmerman and S. Langaas (eds). IWA Publishing, London.

Morrison, K. (2003). Stakeholder involvement in water management: necessity or luxury. *Balancing Competing Water Uses – Present Status and New Prospects. Water Science and Technology* **47**(6), 43–54.

Mostert, E. (2002). The Challenges of Public Participation. *Water Policy*.

Roll, G., Prasad, S., Santos, N., Pulver, M., Roll, M., Talviste, P., Unt, P. and Alyakseeva, N. (2002). *Information and Communication Strategy Report for the Lake Peipsi Basin within the MANTRA-East Information Module*. Center for Transboundary Cooperation, Tartu, 87 pp.

Roll, G. (2003). Generation of usable knowledge in implementation of the European water policy (Chapter 3). In: *Environmental Information in European Transboundary Water Management*, J.G. Timmerman and S. Langaas (eds). IWA Publishing, London.

Rowley, J. (1998). What is information? *Information Services and Use* **18**, 243–254.

Timmerman, J.G., Kipper, K., Meiner, A., Mol, S., Nieuwenhuis, D., Roll, G., Säre, M., Sults, Ü. and Unt, P. (2002). *The use and valuing of environmental information in the decision making process: an experimental study*. Workshop Report, RIZA Report No. 2002.069X RIZA, Lelystad, The Netherlands.

Timmerman, J.G. and Langaas, S. (eds) (2003). *Environmental Information in European Transboundary Water Management*. IWA Publishing, London, 261 pp.

Timmerman, J.G. and Langaas, S. (2004). Water information – what is it good for? On the use of information in transboundary water management. *Regional Environmental Change* **5**(4), 177–187.

Ulland (2001). *Epistemic Communities and International Policy Coordination* by Peter M. Haas. http://www.stanford.edu/~ahm/class/2001-Q1-Win/Soc213-instlThyStates/Ia/Haas1992.rtf

Wittgren, H.B., Westerlund, S. and Castensson, R. (2000). *The Genevad River study. An actor game on the implementation of environmental quality norms for nitrogen in a river basin (in Swedish)*. VASTRA Report No. 1, 265 pp. http://www.vastra.org/publikationer/rapporter/genevad/genevad.pdf (accessed on October 7, 2003).

7

The way ahead for transboundary integrated water management

G.D. Gooch, P. Stålnacke and G. Roll

7.1 INTRODUCTION

Integrated management becomes a particularly complex challenging task when two or more countries share a river and its drainage basin. About 261 transboundary basins covering 45.3% of the earth's continental land surface (excluding Antarctic) have been identified (Wolf *et al.* 1999). These figures illustrate the high relevance of developing appropriate management tools and policy strategies tailored to address transboundary issues in an integrated water resource management (IWRM) system.

In this volume we have shown how integrated qualitative and quantitative scenarios can be developed and formulated as an input into transboundary water management. Some of the major problems facing integrated methodologies have been identified and we have stressed the importance of combining insights from both social and natural sciences. In this concluding chapter we will examine the ways that the scenarios can influence practical water management practices. We will identify some of the problems and discuss possible ways of minimising these problems. Here, building upon the results from the preceding chapters and Gooch

 Integrated Transboundary Water Management in Theory and Practice: Experiences from the New EU Eastern Borders edited by Geoffrey D. Gooch and Per Stålnacke. ISBN: 1843390841. Published by IWA Publishing, London, UK.

(2003c), we will touch upon the vital question of the communication of scientific information and the ways in which existing organisational and institutional structures form and influence the reception of scientific information. We will also discuss the ways in which scientific knowledge must be combined in water management with political and colloquial knowledge, that is, the knowledge held by politicians and other policy-makers, and by the public. Both theoretical and practical recommendations are presented at the end of this chapter.

7.2 PUBLIC, POLITICAL AND ADMINISTRATIVE INFORMATION FOR TRANSBOUNDARY WATER MANAGEMENT

An important aspect of the *MANTRA-East* project was the collection of information concerning public, political and administrative views on transboundary water management. Interviews were conducted with water managers in Estonia and in the Pskov oblast of the Russian Federation in 2003, and focus groups and group discussions were conducted with members of the public and stakeholders during the whole project period 2001–2004. Some findings from these studies were presented in Chapter 3 and detailed results of the Russian interviews and descriptions of environmental decision-making in the Pskov region of the Russian Federation can be found in Kalashnyk (2003) and in Lundqvist (2003). The results of interviews with the public and stakeholders in Estonia can be found on the MANTRA-East website at http://mantraeast.webmedia.ee.

On the ground, the opinions of environmental managers demonstrate that environmental administrators found it difficult to describe the decision-making processes, both of their own countries and especially of their counterparts across the border. They also had difficulty in describing the main ideas behind environmental management. This indicates a lack of institutional transparency, especially in Russia. Hierarchical decision-making processes were common there, although while it was most usual that the heads of department approved all decisions, the specialists also had a reasonable degree of independence to organise their work. The importance of national and European legislation was notable, and the dominating influence on environmental management was seen to be the regulatory legislative structures. The concept of subsidiarity seemed not to include local levels to a significant extent, and at the local level the decision-making process was seen by many (but not all) of the managers as top-down. The problems described by the environmental managers were in some ways similar on both sides of the border. They ranged from lack of resources, lack of coordination of tasks, to the influence of party politics in environmental administration. As transboundary relations were a central focus of the MANTRA-East project, the managers were also asked about transboundary cooperation. Cooperation between local authorities was considered positive, and a number of examples of successful cooperation were described. Relations between the municipal and county authorities were not, however, described as a prioritised issue by the local authorities, although cooperation with the County Environmental Departments in Estonia was good. Many of the managers felt that local–state relations were more important than local–regional relations. Here it was possible to suspect that the inter-organisational competition that was apparent on the Russian side also existed on the Estonian side of the border. Officials at the county level in Estonia also felt that communication with the central level, in the form of the Ministry of the Environment, was too hierarchical. As in Russia, water pollution was considered the major problem in Estonia, although it is interesting to note that the majority of the managers did not consider agriculture as the major problem in this respect. The reasons for this was that many of the managers felt that decreases in agricultural production, as well as the decreases in the use of artificial fertilisers, had led to decreases in the magnitude of the problems of diffuse pollution from farming land. Research within the MANTRA-East project has shown, however, that this was not the case, and that residual effects of fertiliser use are significant.

Transboundary relations between the authorities working in the environmental sectors in Estonia and Russia were, during the beginning of the 2000s, relatively undeveloped, despite the creation of the transboundary water commission. During the Soviet period the border had not had any great significance as far as environmental management was concerned, and exchanges of knowledge and other forms of contact had existed. This changed radically after Estonian independence, and many managers in both Russia and Estonia had little knowledge at the time of the conditions and forms of environmental management on the opposite side of the border (Kalashnyk 2003). Levels of mutual trust had also deteriorated and some of the managers even expressed doubts over the level of interest in environmental issues on the part of their colleagues across the border. In Russia, channels of communication between central, regional and local authorities were erratic and decision-making procedures were difficult to understand (Kalashnyk 2003). Managers on both sides of the border admitted that in certain aspects they had little information on the nature of the other country's environmental problems, although they expected environmental problems to be similar to their own. The majority also believed that the main problem for water quality in Lake Peipsi and its drainage basin was that municipal and industrial wastewater was discharged untreated, or inadequately treated.

Transboundary environmental cooperation between Russia and Estonia seems not have improved during the past 5 years, it may even have deteriorated. A relatively large number of environmental projects involving (and often financed by) third-party countries such as Denmark, Finland, Sweden, etc. have been conducted, and they have tended to overshadow Estonian–Russian cooperation. The reasons for this are many. First, poor finances have been, and still are, a limiting factor for cooperation, and practical problems such as difficulties in crossing the border were acute during the early 2000s (see Chapter 6). Border controls between the regional centres of Tartu (Estonia) and Pskov (Russian Federation) could during this time lead to delays of several hours. Another reason for the limited level of cooperation was that there seemed to have been little consensus among water managers on the most suitable administrative level at which to develop cooperation. Some felt that the state level was the most significant; others felt that projects developed at the local level could be more successful. Some of the managers suggested that more transboundary projects of a general nature (not necessarily environmental or water related) might be a way of improving relations, as knowledge and experiences could be exchanged in such projects. Through these trust and working routines could be developed. Considering the problems of data acquisition from the Russian side of the drainage basin noted in this volume, this could be an important step forward. Staff exchange between water authorities might also be a means of increasing cooperation, although interest in exchange programs has been shown to be asymmetric, with a greater interest in exchange programs being expressed by the Russian authorities. Unfortunately, direct contacts between managers seem to have decreased and have been at quite a low level during the MANTRA-East years. Comprehensive information on transboundary environmental conditions seemed often to have been restricted to a limited number of experts involved in international cooperation.

As noted earlier, Estonian–Russian relations (and quite likely relations along the new European Union (EU) border) in the field of water management appear to be faced by four major problems. The first is the lack of resources, both financial and administrative, of the local environmental authorities; the second is the complicated environmental management systems, especially in Russia, and the lack of understanding of the decision-making processes by the people working in the local environmental authorities; the third is the lack of knowledge of the nature of environmental problems and the means of managing them in a transboundary context; the fourth is the insufficient awareness on the part of some local environmental managers of the nature of problems connected with diffuse pollution from agricultural land. The lack of financial resources is a problem and a major dilemma for many managers working in water management, and although foreign investments and support are to a certain extent available, increased economic help, especially on the Russian side of the border, is necessary. An analysis of the economic problems facing water management is unfortunately outside the scope of this volume. We will discuss the other three major problems in the following sections.

7.3 NATURAL SCIENTIFIC INFORMATION FOR TRANSBOUNDARY WATER MANAGEMENT

Understanding future developments in water quality is an integral part of understanding the consequences of changes in land use, pollution emission and changes due to abatement strategies as imposed by water policies and decisions. Quantitative scenarios and/or 'what-if' assessments may be a vital tool for improved knowledge and important for decision support. For example, many climate scenarios predict a change in the hydrological cycle, comprising less frequent but more intense precipitation. The result will in this case be increased runoff intensities, which again will lead to increased pollution transport from agricultural and urban areas and increased erosion and sediment loads to the rivers, including the load of sediment-associated pollutants. Additionally, a combination of knowledge from observations (surveys and monitoring) and improved process understanding integrated in models is important for addressing these problems. The improvement of existing large-scale river basin models for IWRM is in this respect an important task for the future.

A prerequisite for the successful environmental management of surface and marine waters is the collection of basic environmental statistics and quantitative estimates of the riverine loads, estimation of the pollution sources, retention and buffering capacity in the drainage basin, and knowledge of water quality. The example shown in Chapter 2 has clearly demonstrated that this is far from a trivial task. For example, standard water quality measurements of total nitrogen are almost entirely lacking from the Russian part of the Lake Peipsi drainage basin. The results presented here also imply that, for international or transboundary lakes, strong cooperation in the field of environmental monitoring as well as the exchange of data between the countries sharing the lake and its drainage basin are needed. Even though these tasks may seem trivial, experience worldwide shows that this is not normally the case. Lake Peipsi is one such water basin where existing monitoring programmes suffer from financial limitations and a lack of harmonised monitoring strategies. For instance, the magnitude of the nutrient loads and source emissions has been uncertain for a considerable period of time. While the interconnections between the various terrestrial and aquatic media have received more attention in recent research projects, transboundary basins on the European fringe are often 'data-poor' or 'information-poor' and characterised by highly varying quantities and qualities of input data. As most models require very detailed and spatially consistent input data, their applicability may be limited under such conditions. Thus, simple strategic models or tools are needed to address the limitations in these basins (see Chapter 5). Unfortunately the development of such models is still in its infancy. Another important issue that has to be taken into account is the fact that agriculture in the Eastern Europe has undergone dramatic changes during the past 8–10 years. The years 1990 and 1991 constitute the starting point of an almost total collapse in agricultural production due to the disintegration of the former Soviet Union. The consequences of this collapse were immediate, and included, among other things, dramatic decreases in nutrient inputs. Mineral fertiliser use decreased by 90% and the number of livestock decreased by 50% from previous levels. At the same time, large areas of arable land were taken out of production. Such abrupt and significant land use change has hardly been recorded in modern European agricultural history. The results from our case study show that nitrogen loads have responded rather quickly to these changed emissions while phosphorus loads obviously have not responded to any particular degree (see Chapter 2).

7.4. THE WATER FRAMEWORK DIRECTIVE AS A TOOL FOR TRANSBOUNDARY WATER MANAGEMENT

The EU Water Framework Directive (WFD) (Directive, 2000/60/EC) is probably the most significant legislative instrument in the water field to be introduced at an international level for many years. One of the first steps in the implementation of the WFD (2000/60/EC) is to identify river basins, assign them to river basin districts

(RBDs) and to appoint competent authorities to manage the districts (Article 3). If a river basin extends across international boundaries the directive specifically requires it to be designated to an international RBD. The directive further specifies that countries shall ensure cooperation for producing one single River Basin Management Plan (RBMP) for an international RBD falling within the territories of the EU. However, if this is not possible, the requirement is to produce a RBMP only for the part of the basin falling within each country's respective territory. If the basin extends beyond the territories of the EU the directive encourages Member States to establish cooperation with non-Member States and thus manage the water resource on a basin level (Articles 3 and 13). The guidance document Best Practices in River Basin Management Planning, produced as a part of the Common Implementation Strategy, touches upon international RBDs but does not actually go any further than the directive in specifying how to designate international RBDs.

A study by Nilsson *et al.* (2004) showed that 30% of the prospective RBDs identified under the WFD are international (Figure 7.1). Areawise, the international RBDs constitute about 65% of the total area of prospective RBDs. It is thereby quite clear that for fulfilling the intentions of the WFD regarding management according to river basins, this must imply international management. The directive addresses international river basins in, for example, the designation of RBDs, where it requires international river basins to be assigned to international RBDs. On the other hand,

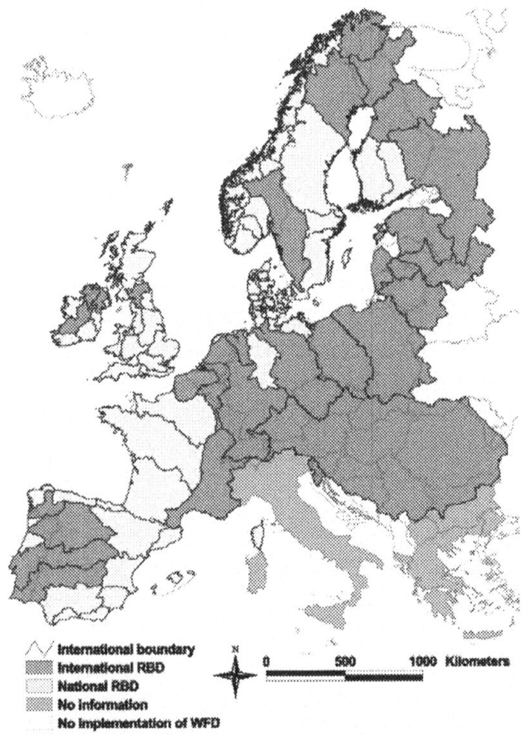

Figure 7.1 Prospective RBDs. After Nilsson *et al.* (2004).

as already stated, the directive is much less strict and more ambiguous in its demands for international RBDs compared to its demands for national districts. Given that such a large proportion of the prospective RBDs cover the territory of more than one country, it might be argued that the 'soft' requirements of the WFD regarding international RBDs may undermine the intentions of the directive of management according to river basins. The directive appears to serve as an incentive for joint management rather than strictly enforcing international management.

Additionally, no common definitions of 'international RBDs' or detailed guidelines exist under the WFD Common Implementation Strategy for the organisation of transboundary collaboration for the implementation the WFD. As a result, transposition into national legislation, and ways of implementing WFD requirements, is taking place in different ways in different countries.

For instance, in the Danube River Basin, that is the second largest in Europe with an area greater than 2000 km^2, 13 countries, the majority of which were not members of the EU, agreed to make the necessary efforts to arrive at a co-coordinated international RBMP for the Danube River Basin. The International Commission for the Protection of the Danube River coordinated the preparation of the international RBMP.

Finland has set up two international RBDs and organised coordination of the implementation of the WFD through transboundary water commissions that are based on agreements with the respective governments of the neighbouring countries. The Torniojoki River International Basin District, for example, involves two EU Member States: Finland and Sweden. The Tenojoki–Paatsjoki international RBDs consists of catchments discharging to the Arctic Ocean, and will require collaboration between Finland, Norway and Russia (see more at http://www.ymparisto.fi/default.asp?node=9812&lan=EN). International RBMPs are also to be developed for rivers Rhine, Elbe, Odra, Scheldt and Meuse.

On the other hand, Estonia, which borders Russia and Latvia, established the Koiva/Gauja international RBDs jointly with Latvia. As a paradox, Estonian authorities have only defined the Estonian side of Lake Peipsi drainage basin as a RBD.

The results from the lake water quality and ecological assessment described in Chapter 5 also demonstrate the difficult, or even impossible, task of finding one universal criterion to diagnose all types of lakes. Each lake type has its own specific features and reacts differently to environmental pressures. Instructions on how to typify lakes and how to estimate their ecological quality ('health') within different types are provided in WFD. Correct and reliable information on the status of water systems is indispensable for rational and cost-effective water management, as demonstrated by the WFD. This has general validity for all types of river basins, but is particularly relevant and problematic for transboundary water regions as clearly demonstrated in Chapter 4. At the same time, monitoring network data and information activities also constitute around 80% of the management activities of the various European transboundary water commissions (R. Enderlein, UN/ECE 1999, personal communication). Given the major political and economic differences as well as variations in data availability, accessibility and quality, this has proven particularly troublesome for many transboundary water regions on the Eastern European fringe such as the Lake Peipsi drainage basin.

7.5 USE OF MODELS IN THE SCENARIOS FOR LAKE PEIPSI

The second step of our scenario methodology involved the quantification of the qualitative scenarios and the use of these quantitative results through computer-based modelling tools. It is here that one of the major challenges to integrated methodology needs to be addressed. The general indications of possible future developments presented in Chapter 4 needed to be translated into figures that could be utilised in the modelling process, and a number of problems were apparent. First, the qualitative scenarios were based on factors and variables, some of which are not easily quantifiable. Political and administrative competence cannot be easily translated into the quantitative modelling process, other than to indicate a state of inertia or activity in pollution prevention, and possible changes in the level and amount of nutrients. The geographical scale of the storylines was also problematic. The qualitative scenarios described possible developments in areas that lacked precise spatial boundaries. A major task for the modellers was therefore to determine the scale of operationalisation, and to define the spatial borders within which the modelling process would occur.

It is usually agreed that the range of model complexity should concur with the nature of the problem and the amount of data available to build and validate the model. Hydrological modelling tools in river basins of different sizes are well-established, with a variety of tools available for different spatial scales and levels of physical base. However, water quality modelling in river basins represents another level of complexity. While the factors and processes that control the emissions and transformations of various pollutants in soils, groundwater and surface waters can be identified and studied relatively well, the heterogeneity, both in terms of spatial and temporal variability, hampers general, regional and large basin estimates. Description of the fate and flux of a pollutant, including the underlying hydrological processes, requires a fairly detailed understanding of (i) the variability in climate and hydrometeorological conditions, (ii) the absolute and relative importance of point and diffuse sources including land use and management practices and (iii) the relative importance of all major hydrobiogeochemical processes involved. Despite these difficulties, modelling of nutrients in catchments has progressed considerably during the last 10–15-year period. Nevertheless, problems still exist, as many models are (i) tailor made for specific purposes, with different ecological and spatio-temporal resolution, (ii) developed for data-rich river basins, or (iii) developed primarily for scientific purposes and the implications for practical management are often neglected or disregarded. Ecological and chemical water quality models in lakes simulate the fate of pollutants and the state of selected water quality variables in water bodies. They incorporate a variety of physical, chemical and biological processes that control the transport and transformation of these variables. Water quality models are driven by hydrodynamics, point and non-point source loadings, and key environmental forcing functions such as temperature, solar radiation, wind speed, pH and light attenuation coefficients. The external drivers may be specified from observed databases, or simulated using specialised models describing, for example, the water body hydrodynamics or the watershed pollutant loading. The internal forcing functions may also be specified from databases, or calculated within the water quality model using a range of empirical to mechanistic process formulations. Examples include temperature, pH and light attenuation.

For decision-makers and scientists faced with a specific water management problem it is therefore essential to choose methods that provide the expected results, and that are of course at the same time economically feasible. The methods must also take into account data availability, which is often limited. Unfortunately, comparable studies of different methods and models are scarce, and model applications are often performed and described for specific watersheds where other models have not been applied. Continuous changes in land use and agricultural practices also prevent comparison when models are applied in the same watersheds, but on different occasions. Moreover, results from a comprehensive pan-European modelling exercise (in the EURO-HARP project; www.euroharp.org) clearly indicate that the collection and organisation of the input data for the various models is an extremely time-consuming task, even in study areas where data are already available. It can be expected that this could be an even more difficult problem for transboundary basins.

Spatial and time scales are the key elements of an integrated modelling approach at the level where integrated strategic planning and water management is conducted, since long-term effects have to be considered in the decision-making processes. Therefore, an integrated modelling tool at the river basin scale has to consider these large ranges for temporal and spatial scales. It also means that all subsystems that have a significant contribution to the water management problems under consideration have to be included. Although our scenario-modelling exercise is marred with uncertainty especially for small-scale predictions, our results also provide very satisfying results at the scale of the entire Lake Peipsi drainage basin. The simulated loads are in agreement with previous estimates, and the massive decrease in nutrient loads during the 1990s can be simulated quite well. One of the most important advantages of the approach is that nutrient losses can also be estimated for data-poor areas and ungauged basins. This is an important result in a transboundary context. The situation in other drainage basins around the so-called 'Eastern European fringe' is comparable with the border area between Estonia/Latvia and Russia, where, for the Russian part there is often a lack of data, and even if data is available it can be very difficult to obtain (see Chapter 2). Using conclusions and emission estimates from data-rich areas, satisfactory load simulations can be obtained for the other areas by using simple easy to obtain geographical information system (GIS) data layers like land use and river networks.

The forward-looking nature of the modelling makes it very suitable for quantitative assessments of future scenario storylines. Although the simulated loads into the lake under the various scenarios that were defined for this area cannot yet be checked against observed loads, they reveal considerable information on the sensitivity of the nutrient loads to changes in emissions. Source apportionment, both spatially ('regions') and thematically (diffuse and point sources) make it possible to set priorities for management.

7.6 INSTITUTIONAL FRAMEWORKS AND WATER POLICY

In this volume we have stressed that it is important to realise that water management takes place within institutional contexts and that these contexts determine the ways in which scientific information, such as that produced by the integrated scenarios, is received and used. Information needs to be translated into knowledge that can be utilised in management and policy processes.

How then to improve the communication of knowledge, and what are the barriers to communication? In the case of Lake Peipsi the lack of administrative resources and the existence of opaque decision-making processes on the Russian side, together with a lack of understanding of institutional processes all pose major problems (see Chapter 3). The existence of different administrative water management systems also complicates matters. Estonia was at the time of the MANTRA-East project developing management structures influenced by EU concepts of subsidiarity and multilevel governance, while in Russia former soviet management systems continued to be used, based on top-down, hierarchical institutional interaction. The strongly centralised Russian system of environmental management complicated potential forms of transboundary cooperation, and this problem has persisted.

Due to these complications it is vital to analyse the role of institutional cultures in the transboundary river or lake basin. Of special interest are the relationships between formal (organisational) and informal (norms and values) institutions. The MANTRA-East project further developed and utilised the institutional framework presented in Table 7.1, which builds on Pfeffer (1981). In what is of course a simplification of complex factors three dominant institutional models were formulated from the interviews with water managers. These models were the Rational Model, a normative ideal for many water managers, the Bureaucratic Model and the Political Model. Table 7.1 presents a number of aspects of the models such as goal and preferences, forms of power, decision processes, beliefs about action and consequences, forms of decision-making, ideologies and organisational forms. Of special interest was the different ways that scientific information could be used.

Scientific information in the Rational Model is seen as a major determinant of the decision process. It is expected to be extensive and systematic. A major problem here is, as has been noted in a number of contributions in this volume, that access to information can, and often is, limited. Also, it may not always be presented in an easily accessible way to decision-makers (see Chapter 6). The costs of obtaining complete information, even if it was possible, would also be exorbitant. In practice this means that the use of information, even in the normative Rational Model, is limited, and that many authorities are by necessity forced to accept not complete, but *satisfactory* levels of information. A state of *satisficing* becomes the norm.

In the Bureaucratic Model information use is dependent on established procedures. Information that fits into these procedures will be able to be utilised, information that does not falls outside the established rules of use. In practice this means that conditions of incomplete information use exist even here.

Table 7.1 Institutional contexts in transboundary water management (Gooch).

Dimension	Rational	Bureaucratic	Political
Goals/preferences	Consistent and based on complete information	Determined by organisational rules	Pluralistic, formed by competing groups
Power	Centralised and transparent	Rule determined and not always understandable by outsiders	Determined by coalitions that can change
Decision process	Orderly, attempts to mirror economic principles	Procedural, follows established patterns	Disorderly and incremental. The result of conflicts
Information use	Extensive and systematic	Used only when it fits into established rules and procedures	Used strategically by different coalitions
Beliefs about action consequences	Based on known probability; predictable	Expected outcomes based on earlier experiences	Disagreement on outcomes between coalitions
Decisions	Perceived as value maximising	The result of established routines	The result of bargaining between coalitions
Dominant ideology	Economic efficiency	Stability and predictability	Conflict as a creative force
Organisation	Market-based, New Public Management	Hierarchical	Competing networks
Examples of theory	Institutional Rational Choice	Max Weber's theory of bureaucracy; cultural theory	Advocacy Coalition Framework
Theorists	Elinor Oström	Max Weber, Aron Wildavsky	Paul Sabatier

In the Political Model scientific information is used as a weapon between competing coalitions and groups. The value of information is therefore context dependent, and different groups are likely to use different interpretations of similar information sources, or simply utilise completely different sources of information. Information in this context also becomes a valuable resource that can be withheld or traded for the benefit of the coalition.

Looking once again at the context of transboundary water management in the Lake Peipsi region we could see that the use of information on the Russian side of the border could often be characterised as an example of the way that information is used in the Political Model. Authorities at various levels may have seen information as a valuable asset that should not be shared without clear benefits in return. This approach contrasts of course with any other authority that views that use of information from the Rational Model's perspective.

The consequence of the use of information in the different models becomes problematic when two or more organisations need to cooperate in, for example, a transboundary context. It is therefore important to analyse the institutional contexts of the participating organisations and authorities, and to develop an information strategy that is acceptable to all partners.

7.7 CONCLUSIONS AND RECOMMENDATIONS

IWRM is a term often used without further and deeper consideration (Biswas 2004; Allan 2005). Far from being an accepted and easily understood notion around which there is consensus, the entire concept and exact definition of integrated water management (IWM), IWRM and integrated river basin management (IRBM) are in fact widely debated. The roots of IWRM can be traced back to the establishment of the Tennessee Valley Authority (TVA) in 1933. The TVA effort targeted many of the elements included in modern IWRM perceptions: natural resource utilisation planning combined with sustainability principles on balance between social, economic and environmental objectives.

The more modern concept of IWRM is primarily conceived for the purpose of promoting sustainable water management. In 1977, at the International Water Conference in Mar del Plata, the emphasis was on coordination between water sectors primarily at a national authority level (Snellen and Schrevel 2004). The need for sector coordination and holistic approaches was given further attention at the 1992 Dublin conference, preparing for the Earth Summit in Rio de Janeiro. Later important milestones were the Millennium Development Goals 2000 and the Johannesburg conference (UN 2002). These conferences advocated that IWRM should go further than just including the coordination of sectors and agencies and interaction between ground and surface water, but also include the carrying capacity of the natural environment and demand management (Koudstaal *et al.* 1992). The concept is however often ambiguous, and a unanimously agreed definition of it has yet to appear. In an attempt to remedy this, the Technical Advisory Committee (TEC) of Global Water Partnership (GWP) has adopted the following definition:

> IWRM is a process, which promotes the co-ordinated development and management of water, land and related resources, in order to maximise the resultant economic and social welfare in an equitable manner without compromising the sustainability of vital ecosystems (GWP-TEC 2000).

Key points here are process, coordinated, and the important relationship between sustainability and economic and social welfare. The definition does not, however, provide us with much indication of how this coordination (and integration) is to be achieved, as pointed out by Gooch and Stålnacke (2003). Recently other authors have also argued that the IWRM approach as it is defined by the GWP, cannot be implemented in practice generally due to operational questions and related problems of establishing measurable criteria (e.g. Biswas 2004; Jeffrey and Gearey 2005; Lankford and Cour 2005). Among the barriers to implementation mentioned is the need for more centralised policy development and implementation and thus larger, slower and more bureaucratic authorities to handle policy aspects. Furthermore, objectives like stakeholder participation and decentralisation would be unlikely to promote integration. Recently Pahl-Wostl and Sendzimir (2005, unpublished) summarised the following main limitations of IWRM:

- The vagueness of the concept.
- The integration of many topics and management of a system for a specific purpose cannot be realised simultaneously.
- The claims for change in management practices made by the promoters of IWRM seem to be based on normative claims rather than on a sound scientific base that would provide evidence of these new management approaches.
- Examples for the successful implementation of IWRM are lacking.
- The emphasis on process without clearly defined and measurable targets for the goals to be achieved.

Rahaman *et al.* (2003) analysed the WFD principles in relation to IWRM (especially for developing countries) and identified seven mismatches:

(1) Gender awareness is omitted. No guidelines for women's role in the provision, management and safeguarding of water.
(2) Decentralisation is neglected. No guidelines to encourage and regulate the private sector.
(3) No guidelines to ensure coordination between different sectors.
(4) The EU WFD does not set up clear guidelines for the active participation of local people and water users in the management of water.
(5) No focus on poverty.
(6) The EU WFD promotes technology-oriented management in the case of drinking water and sanitation, and as a way of ensuring a good quality of water. But better management calls for the integration of technology-oriented management with human-oriented management.
(7) No standard guideline to develop responsibilities to the lowest appropriate level.

In an attempt to alleviate some of these problems the approach presented in this volume was based on the creation of alternative water management scenarios that combined input from various scientific disciplines. The procedure included the following sequential steps:

(1) The creation of mutual disciplinary respect and a basic understanding of all scientific approaches.
(2) The definition of three to four qualitative storyline scenarios on regional development.
(3) The translation of the qualitative storyline scenarios into quantitative GIS layers.
(4) The modelling of the nutrient fluxes and loads in the drainage basin.
(5) The modelling of the transformation of nutrient in the lake and the ecological consequences.
(6) Analyses of the policy and socio-economic implications of the modelling results, including an analysis of the value of scenarios from an information perspective.

Our experiences have demonstrated that an approach can be developed that respects the integrity of individual scientific disciplines while at the same time combining and integrating approaches so that synergic results are obtained. The first basic step is to create a working atmosphere that combines mutual respect with scientific questioning; that is, representatives of different disciplines must learn enough of each other's approaches to be able to dispute both the methodology used and the results obtained. While this critical questioning will by necessity be restricted to a non-expert level, it is still important that research team members do not simply accept other disciplinary paradigms at face value. The use of scenarios has enabled participants in the MANTRA-East project to build upon the knowledge and results of both natural and social science approaches, and to combine these in a reflective and alternate exchange in which qualitative and quantitative methodologies are combined.

7.8 PRACTICAL RECOMMENDATIONS

A major aim of the project was to provide both theoretical and practical recommendations for integrated transboundary water management. The major practical results and recommendations are:

- *There is a need to conduct comparative policy science studies to promote efficient water management planning*. A review of existing transboundary water management structures and practices in transboundary water basins in Europe demonstrated a lack of well-developed research and analysis on the implementation of water management policies and plans (Gooch *et al.* 2002). It was also shown that organisational and institutional aspects of implementing EU water policy (political, research, administration, etc.) need to be developed. Problems of communication and information exchange between different levels of governance as well as across borders present major difficulties for water policy implementation.
- *The political processes of transboundary cooperation must be taken into account*. Water management in a transboundary context is much more complex than water management within one state. There is not one government to manage the transboundary waters, there are different states with their distinct political and economic interests, different histories and cultures (see Chapter 6). The management of transboundary water is inherently political, and political will from the governments of all riparian countries is a prerequisite for a successful initiation and continuation of any transboundary cooperation. Actions should be taken to promote political commitment to international cooperation by the transboundary states.
- *There is a need to establish formal arrangements and detailed procedures for transboundary cooperation*. Formal arrangement and procedures should be established between the riparian governments as well as between the governments and the stakeholders. Their responsibilities and procedures of work should be clearly described, especially in those aspects that concern the implementation of the WFD in transboundary waters.
- *There is a need to address differences in water management competences between countries*. Investigations during the MANTRA-East project illustrated that there is a growing gap in the formal frameworks (different administrative

structures, norms and standards), practices, information and levels of funding towards water protection measures on the different sides of the Estonian–Russian (Gooch 2003a). This presents a major challenge for the trans-national implementation of EU water initiatives such as the WFD.

- *Stakeholders and the public participation are the keys to the successful implementation of water policies.* The review of the experiences of stakeholder and public participation in transboundary water management confirmed that involving multiple stakeholder groups in the development and implementation of EU and national water policies is critically important. However, this is not always feasible for various reasons. Usually, there are in transboundary water basins only a few organised stakeholder groups that are in some way involved in the planning and implementation of water policies. Many local stakeholders are not sufficiently aware of regional water management issues and therefore are not able to become involved. Traditionally, a major bottleneck in the implementation of water policies is created when experts produce a highly technical body of information that becomes incomprehensible to non-experts. Innovative approaches and technologies to disseminate water management information (e.g. semantic webs, citizen juries) were tested and found to be valuable in transboundary water basins as a means of increasing awareness (Säre and Roll 2004). An important tool in the Lake Peipsi Basin was the regional Internet portal (www.peipsi.org). This used knowledge management technological solutions to provide interested parties with comprehensive information and news on environmental and regional development issues in the lake basin. The Lake Peipsi portal was available in Estonian, English and Russian as it is important to use the mother tongue of local stakeholders for this kind of communication tools.
- *Transboundary Water Commissions need to involve stakeholders.* Seven case studies examining various aspects of environmental information use from both theoretical and empirical perspectives were conducted for the Nemunas River, Bug River, Oder River, Lake Neusiedl, Lake Constance, Elbe River and Spanish–Portuguese Rivers (Nilsson and Langaas 2003). The results show that transboundary water commissions are largely expert/technical in nature. The socio-economic connotations of water management decisions may as a consequence be underestimated. One consequence of the expert/technical of commissions can be a lack of attention to the involvement of stakeholders.
- *Socio-economic information is important in the water management decision-making process.* Environmental data is rarely used in the decision-making process unless it shows a direct and clear connection between the impact of the physico-chemical and biological conditions to changes in the economic and social situation in a given transboundary water region (Timmerman *et al.* 2002). Information for decision-making, especially the analysis of the problem, needs to fall within the scope of expectations of the decision-makers. For a transboundary water management situation this implies that, to be effective, an existing problem should be described from the viewpoints of the countries involved. Furthermore, the information should also allow for different solutions in the different countries (Timmerman *et al.* 2002).

- *The use of environmental and socio-economic information is crucial for transboundary water management.* A very wide spectrum of information is required to support decision-making and to evaluate the effects of water resources management decisions. Within the MANTRA-East project it was found that information production lags behind information needs in water management. Although IWM was introduced more than a decade ago, information about transboundary water basins still focuses mostly on hydrological and ecological components of water bodies and largely ignores the importance of socio-economic data and processes. Among the reasons that hinder production of improved information are: (1) strong boundaries between different disciplines that are not easily overcome; (2) the variety of information needs are underestimated and the knowledge and perception of goals of information dissemination prior to producing the information is insufficient; and (3) differences in institutional behaviour between representatives of different organisations involved in the cooperation hinder the collaboration between these institutions.
- *There is a need to develop innovative approaches and technologies to disseminate water management information.* Semantic webs, focus groups and citizen juries were tested and found to be valuable for the implementation of transboundary water management as they increase the awareness of local stakeholders (Säre and Roll 2004).
- *Simplistic nutrient modelling tools can be efficient for pollution prevention strategies.* In transboundary river basins, riverine load modelling and source apportionment is more difficult than in other situations because the required administrative statistics and GIS (spatial) data are not harmonious for each country. This was especially the case for the Lake Peipsi basin, which can be regarded as data rich in the Estonian part, and data poor with respect to the Russian part. In the project, two models (MESAW and Polflow) were applied to assess the source: retention and transport of nutrients (Mourad and van der Perk 2004; Vassiljev and Stålnacke 2005). Both models were proved to be complementary and useful tools for the assessment of nutrient loads in past, present and future in transboundary drainage basins. With a minimum of large-scale maps and calibration of the model using data from a relatively data-rich part (Estonia), plausible nutrient emission estimations and load simulations could be obtained for an entire basin, including data-poor parts (Russia/Latvia). The modelling of nutrient emissions and loads for future scenarios enables decision-makers to identify priorities for water management, and evaluate the effect of various developments (see next point).
- *Blue-green algal blooms were more frequent in Lake Peipsi due to lowered N/P ratio.* There was a clear decrease in the N/P ratios, at least in the Estonian rivers (Iital *et al.* 2004; Stålnacke *et al.* 2004), which was most likely the reason for increased observations of cyanobacterial blooms in the lake in recent years. More precisely, changed relations in the amounts of N and P in riverine loadings have led to the situation that N/P ratios, both in rivers and in Lake Peipsi, has fallen below the critical ratio of 30 (Nõges *et al.* 2003a). Besides the N/P ratio, the concentration of phosphorus is an important factor determining cyanobacterial development. At high P levels and low N/P ratios bloom forming

blue-greens are favoured due to their ability to fix atmospheric N_2 in the conditions of N limitation.

- *Model results show the importance of decreased phosphorus loadings.* Model calculations show that decreased phosphorus loads will decrease the risk of water blooms in Lake Peipsi (Nõges *et al.* 2004). As the main proportion of phosphorus was coming through the two major rivers Velikaya (Russia) and Emajõgi (Estonia), main attention should be paid to these two rivers, especially concerning the coordination of actions. Connection to wastewater treatment plants and improved P removal will give an immediate decrease especially for point sources in close proximity to the lake (e.g. Pskov city in Russia) and solve hygienic problems locally, but long-term future strategies for nutrient load reduction should mainly focus on not increasing the agricultural nutrient runoff. Today, agriculture is responsible for the largest portion of the total loading to the lake (Stålnacke *et al.* 2002; Mourad and van der Perk 2004; Vassiljev and Stålnacke 2005) even though the losses from agriculture per unit area (e.g. kg/km^2) is at a very low international level (Stålnacke and Roll 2002; Stålnacke *et al.* 2002). So the future loadings will depend heavily on how agricultural land will be used in future, for example, how much of the present set aside and abandoned land will be used in future and the intensity in the agricultural sector.
- *The Lake Peipsi scenario results show the importance of controlling both point and diffuse sources.* In the MANTRA-East project, qualitative scenarios, 'story-lines' or 'scripts' were developed (Gooch 2003b) and presented in the paper by Gooch (2004). The qualitative/explorative scenarios were then translated into quantitative GIS layers. A nutrient transport model was then used to calculate the nutrient emissions, as well as transport and retention and the resulting nutrient loads into the lake (Mourad *et al.* 2003). These estimated nutrient loads were used to assess the ecological effect in Lake Peipsi (Nõges *et al.* 2003b). The model results from the scenarios suggest that change in the amount of arable land will be the major factor controlling the future nutrient

Figure 7.2 River Emajõgi, the second largest river in the Lake Peipsi drainage basin (9740 km^2). (Photo: Peeter Unt (Peipsi CTC).)

loads to Lake Peipsi. The results also show that the riverine loads of nutrient loads will change surprisingly little even under extreme future changes.

- *There is a need for better cooperation and coordination across borders*. To improve transboundary cooperation it is necessary to initiate actions such as common monitoring programs, the construction of common data bases, common tools for hydrological and ecological predictions as well as efficient information dissemination and exchange systems (Stålnacke *et al.* 2005).
- *There are weak requirements for transboundary RBDs in the WFD*. As argued in Section 7.4 (Nilsson *et al.* 2004) transboundary aspects seem to be seriously underestimated in WFD implementation. Given that many of the RBDs will be transboundary, together with the 'softly' defined requirements of the WFD text concerning the transboundary RBDs (Articles 3 and 13 in WFD), the directive's requirements and high ambitions of holistic management and administration according to river basins may be undermined. Formal and more legally binding arrangements and procedures should be established between the riparian governments as well as between the governments and the stakeholders. Results from R&D projects such as the MANTRA-East project have led to increased knowledge about various aspects of transboundary water- and pollution-related problems. Overall, the activities have clearly demonstrated that the implementation of the WFD is faced with a difficult task in transboundary RBDs especially those located on the new EU border and beyond.
- *There is a need to finance water protection measures*. The implementation of water protection measures requires considerable financial resources, usually much higher than are usually available in transboundary water basins. In this context the environmental objectives of water management plans should be coordinated with the economic development priorities of the border regions.

REFERENCES

Allan, T. (2005). *Personal communication*. Stockholm Water Week, August 2005.

Biswas, A.K. (2004). Integrated water resources management: a reassessment. *Water International* **29**(2), 248–256.

EC (2000). Directive 2000/60/EC of the European Parliament and of the Council of 23 October 2000 – establishing a framework for Community action in the field of water policy. *Official Journal of the European Communities*.

Gooch, G.D., Höglund, P., Roll, G., Lopman, E. and Aliakseyeva, N. (2002). *Review of existing structures, models and practices for transboundary water management*. MANTRA-East Report D1a.

Gooch, G.D. (2003a). The EU, Russia, and the applicant states: water management and institutional interaction on the New European border. *Proceedings from 7th International Water Association (IWA) Conference on Diffuse Pollution and Basin Management (DipCon)*, August 17–22, University College Dublin.

Gooch, G.D. (2003b). *Scenarios for environmental policy-making on the Estonian–Russian border*. Department of Management and Economics, Linköpings University, Linköping, 17.

Gooch, G.D. (2003c). The communication of scientific information in institutional contexts: the specific case of transboundary water management in Europe (Chapter 2). In: *The Role and Use of Environmental Information in European Transboundary River Basin Management*, J.G. Timmerman and S. Langaas (eds). IWA Publishing, London.

Gooch, G.D. (2004). The use of integrated scenarios in transboundary water and river basin management. *Proceedings from Conference on Integrated Water Management of Transboundary Catchments: A Contribution from TRANSCAT*, Palazzo Zorzi, Venice, Italy, March 24–26.

GWP-TEC (Global Water Partnership – Technical Advisory Committee) (2000). *Integrated water resources management.* TAC Background Papers No. 4, GWP, Stockholm, Sweden.

Iital, A., Stålnacke, P., Deelstra, J., Loigu, E. and Pihlak, M. (2005). Effects of large-scale changes in emissions on nutrient concentrations in Estonian Rivers in the Lake Peipsi drainage basin. *Journal of Hydrology* **304**, 261–273.

Jeffrey, P. and Gearey, M. (2005). Integrated water resources management: lost on the road from ambition to realisation? Prepared for *the WATERMATEX Conference*, November 2004, Beijing, China (in press).

Kalashnyk, L. (2003). *Environmental decision-making in the Pskov Region of the Russian Federation.* Department of Management and Economics, Linköping University, Linköping, 101.

Koudstaal, R., Rijsberman, F.R. and Savenije, H. (1992). Water and sustainable development. *International Conference on Water and the Environment.* Key Note Papers by ICWE Secreteriat, c/o World Meteorological Organisation, Geneva, Switzerland.

Lankford, B.A. and Cour, J. (2005). From integrated to adaptive: a new framework for water resources management of river basins. *Proceedings of the East Africa River Basin Management Conference*, Morogoro, Tanzania, March 7–9.

Lundqvist, A. (2003). *Contemporary Russian environmental policy: problems, players and priorities. The case of Pskov Region and environmental agenda-setting process.* Department of Management and Economics, Linköpings University, Linköping, 103.

Mourad, D.S.J., van der Perk, M., Gooch, G.D., Loigu, E., Piirimäe, K. and Stålnacke, P. (2003). GIS-based quantification of future nutrient loads into Lake Peipsi/Chudskoe using qualitative regional development scenarios. *Proceedings of the 7th International Specialised Conference on Diffuse Pollution and Basin Management*, Dublin, Ireland, August 17–22.

Mourad, D.S.J. and van der Perk, M. (2004). Modelling nutrient fluxes from diffuse and point emissions to river loads: the Estonian part of the transboundary Lake Peipsi/Chudskoe drainage basin (Russia/Estonia/Latvia). *Water Science and Technology* **49**, 21–28.

Nilsson, S. and Langaas, S. (2003). A comparative study of environmental information management in three transboundary water regimes (Chapter 16). In: *The Role and Use of Environmental Information in European Transboundary River Basin Management*, J.G. Timmerman and S. Langaas (eds). IWA Publishing, London.

Nilsson, S., Langaas, S. and Hannerz, F. (2004). International River Basin Districts under the EU Water Framework Directive: identification and planned cooperation. *European Water Management Online*, No. 2004/02. http://www.ewaonline.de/journal/online.htm

Nõges, T., Blinova, I., Jastremski, V., Laugaste, R., Loigu, E., Skakalski, B. and Tõnno, I. (2003a). Reduced nitrogen loading enhance cyanobacterial blooms in Lake Peipsi. In: *Proceedings of 2nd International Conference on Sustainable Management of Transboundary Waters in Europe*, F. Bernardini, M. Landsberg-Ucziwek, S. Haunia, M. Adriaanse and R. Enderlein (eds). Miedzydroje, Poland, pp. 501–504.

Nõges, T. (ed.), Pihlak, M., Nõges, P., Vetemaa, M., Loigu, E., Laugaste, R., Skakalski, B., Stålnacke, P., Gooch, G. and Mourad, D. (2003b). *Final report on the results of the potential influence of different development scenarios of Lake Peipsi region on water quality, trophic status and the situation of fish stocks.* MANTRA-East Deliverable D6b, 53 pp.

Nõges, T., Pihlak, M., Vetemaa, M. and Nõges, P. (2004). Potential influence of different loading scenarios of Lake Peipsi (Estonia/Russia) on water quality, trophic status and the situation of fish stocks. *Proceedings from Conference on Integrated Water Management of Transboundary Catchments: A Contribution from TRANSCAT*, Palazzo Zorzi, Venice, Italy, March 24–26.

Pahl-Wostl, C. and Sendzimir, J. (2005). The relationship between IWRM and adaptive management. *Discussion Input for NeWater International Platform at Stockholm Water Week*, August 2005.

Pfeffer, J. (1981). *Power in Organisations*. Pitman Publishing, Marshfield, MA.

Pihlak, M., Laugaste, R., Loigu, E., Nõges, P., Skakalski, B. and Nõges, T. (2003). *Model of Lake Peipsi Ecosystem – SHALMOD*. MANTRA-East Deliverable D6a, 21 pp.

Rahaman, M.M., Varis, O. and Kajander, T. (2004). EU water framework directive vs. integrated water resources management: the seven mismatches. *Water Resources Development* **20**(4), 565–575.

Snellen, W.B. and Schrevel, A. (2004). IWRM: for sustainable use of water – 50 years of experience with the concept of integrated water management. *Background document to FAO/Netherlands Conference on Water for Food and Ecosystems*. http://www.fao.org/ag/wfe2005/docs/IWRM_Background.pdf.

Säre, M. and Roll, G. (2004). Empowerment of public participation in management of transboundary waters in countries in transition: Lake Peipsi (Estonia–Russia) case study. *Proceedings from Conference on Integrated Water Management of Transboundary Catchments: A Contribution from TRANSCAT*, Palazzo Zorzi, Venice, Italy, March 24–26.

Stålnacke, P. and Roll. G. (2002). Lake Peipsi: a transboundary lake on the future border of the European Union. In: *The NEBI Yearbook 2001/2002*, L. Hedegaard and Lindström (eds). Springer-Verlag, Berlin, Heidelberg, pp. 159–176.

Stålnacke, P., Sülts, U., Vasiliev, A., Skakalsky, B., Botina, A., Roll, G., Pachel, K. and Maltsman, T. (2002). An assessment of riverine loads of nutrients to Lake Peipsi, 1995–1998. *Large Rivers* **13**(3–4) (*Archiv Fur Hydrobiologie. Suppl.* **141**(3–4)), 437–457.

Stålnacke, P., Vandsemb, S.M., Vassiljev, A., Grimvall, A. and Jolankai, G. (2004). Changes in nutrient levels in some Eastern European rivers in response to large-scale changes in agriculture. *Water Science and Technology* **49**, 29–36.

Stålnacke, P., Loigu, E., Melnik, M., Nõges, T. and Vetemaa, M. (2005). *Transboundary diagnostic analysis of Lake Peipsi/Chudskoe*. NIVA Report No. 4956-2005, 62 pp.

Timmerman, J.G., Kipper, K., Meiner, A., Mol, S., Nieuwenhuis, D., Roll, G., Säre, M., Sults, Ü. and Unt, P. (2002). *The use and valuing of environmental information in the decision making process: an experimental study*. Workshop Working Paper. RIZA Report No. 2002.069X. RIZA, Lelystad, The Netherlands.

UN (2002). *Plan of Implementation of the World Summit on Sustainable Development*. http://www.un.org/esa/sustdev/documents/WSSD_POI_PD/English/WSSD_PlanImpl.pdf

Vassiljev, A. and Stålnacke, P. (2005). Statistical modelling of rivering nutrient sources and retention in the lake Peipsi drainage basin. *Water Science and Technology* **51**, 309–317.

Wolf, A.T., Natharius, J.A., Danielson, J.J., Ward, B.S. and Pender, J.K. (1999). International river basins of the world. *International Journal of Water Resources Development* **15**(4).

Index

Page numbers in *italics* denote figures. Page numbers in **bold** denote tables.

Aarhus Convention 140–1
administrative systems
  Baltic States 11–13
  EU–Russian cooperation 7
  information 150–2
  Lake Peipsi region 55–82
  legislation implementation 4
  Pskov Region 61–5
  qualitative scenarios 87–92, 96–7
Agenda 2000 74, 75
agricultural sector
  Baltic States 13–15
  Common Agricultural Policy 70–80
  end-user information 138
  Lake Peipsi basin 32–4
  policy process 76–80
  pollution 7, 64–5, 70–2
  scenarios 86–8, 90–4, 125
air pollution 64, 70
algae *see* blue-green algae
associations of local authorities 69
auction system (fishing rights) 95

Baltic States 6–15, 56, 142
baseline scenarios 83–8
benthic food webs 45
blue-green algae 40, *41*, 48–9, *119–20*, *122–3*, 164–5
border regions (EU) 6–8, 131–3
  see also Russian–Estonian border
Bureaucratic Model 158, **159**
'business as usual' scenarios 83–8

calibration of models 108–10
CAP *see* Common Agricultural Policy
cause–effect chain 136–9
CEC *see* Central European Countries
Central and Eastern European Countries (CEEC) 75–6, 79–80, 94
Central European Countries (CEC) *6*, 9
Centre for Transboundary Cooperation (CTC) 139, 144–5
chlorophyll concentrations *40*
citizen juries 145
clear water quality indicator 50
closed systems 19
Committee of Land Resources 61
Committee of Natural Resources 59, 61
commodification of information 135
Common Agricultural Policy (CAP) 7, 70–80
communication issues 127–48
community information 144
Community Preference principle 73
comparative policy studies 162
Confederation of Professional Agricultural Organisations 77
cooperation levels
  Baltic States 11–13
  communication issues 128–30, 132
  environmental management 80–1
  EU external borders 132

cooperation levels (*cont'd*)
EU-Russia 7, 9
information 150–2
inter-municipal 69
Northeast Europe 8–9, 13
practical recommendations 162, 166
qualitative scenarios 85
country scenarios 84
county environmental departments 67, 68
crisis scenarios 85–6, 90–3, 121–2
'critical' view, social sciences 19
CTC *see* Centre for Transboundary Cooperation
cultural connections 128
cyanobacteria *see* blue-green algae

data
information as 134
modelling input 105–6
decentralisation 8–9
decision-making 76–8, 135–9, 144, 150, 163
DEM *see* digital elevation model
diatom analyses 42–3, *43*, *119–20*, *122–3*
diffuse pollution sources 16, 32–4, 104, 107–8
digital elevation model (DEM) 102
disciplinary differences 18–20
DPSIR-framework 136–9
drainage basins 124, 157
see also Lake Peipsi
driving forces **112–13**, 136

Eastern European countries *6*, 32, **33**, 75–6, 79–80, 94
ecological models 116–22
economic development 34, 85, 87, 89–92, 95–6
see also socio-economic development
Emajõgi River 108–9, *109*, *165*
end-user environmental information 133–9
environmental information 133–9, 140
environmental management
Estonia 65–72, 80–1, 150–2
Lake Peipsi region 55–82
Pskov Region 59–65, *63*
Russia 57–65, *60*, 80–1, 150–2
scenario concepts 96–8
scientific information 152–3
environmental problems 13–15, 70
epistemic communities 129
Estonia
administration/politics 56
environmental management 65–72, 80–1
ethnic relations 10–11
grass/wetlands *94*
Lake Peipsi *30–1*, 32, 35–40, **38**, *39*
local government 67–70
see also Russian–Estonian border
Estonian National Strategy (NES) 66
ethnic relations 10–11, 90
European Union (EU)
border regions 6–8, 131–3
Common Agricultural Policy 70–80
cooperation problems 7, 8–9
external borders 131–3
integrated approach need 16
policy changes 1–4
transboundary river basins 4, *5*
Water Framework Directive 153–5
eutrophication 42–3, 146
external borders (EU) 131–3

fast development scenarios 85–6, 88–9
F/C scenarios 92–3
fertiliser use 14–15, 32, **33**
financial organisation
CAP 71, 75, 78, 80–1
communication issues 131
Estonia 67–9
practical recommendations 166
Russian regions 57, 65
Financial Solidarity principle 73, 78–9
fish-kills 45–7, *47*, 49–50
fish stocks
Lake Peipsi 29, 42, 44, **45**, 51, 64
qualitative scenarios 87, 88–9, 92
fishing rights 95
food chains 44–5, *46*
France 79
funding *see* financial organisation

geo-politics 9–11
geographical features, Lake Peipsi 26–9
geographical information system (GIS) 102
Germany 79–80
GIS *see* geographical information system

Global Water Partnership (GWP) 160
governance 76, 78
grass/wetlands *94*
gross emission calculations 111
groundwater systems 104–5
GWP *see* Global Water Partnership

hydrological processes 34, *35*
hydrological sub-model 104, *123*

ice-fishing *166*
ICS *see* information and communication strategy
impacts, cause–effect chain 136
incremental policy implementation 13
information
  definitions 134–5, *135*
  end users 133–9
  needs and channels 143–7
  water management 150–3, 158–60, 162–4
information and communication strategy (ICS) 139–46
institutional frameworks 130, 158–60, **159**
integrated approach
  methodology 21–2
  need 15–16
  problem identification/solving 1–24
integrated river basin management (IRBM) 16–17
integrated water management (IWM) 16–17
integrated water resource management (IWRM)
  definition 16–17, 160
  future directions 149–68
integration
  definition 16–18
  forms **18**
inter-municipal cooperation 69
international communications 130
international river basins *5*, 154–5
Internet use 143–6, *145*
'interpretive' view, social sciences 19
intervention scenarios 84
IRBM *see* integrated river basin management
isolationist scenarios 85–6, 91–2
IWM *see* integrated water management
IWRM *see* integrated water resource management

joint bodies 130

knowledge, information as 134

labile phosphorus 50
Lake Chudskoe *36*
Lake Lämmijärv 28
Lake Peipsi 16, *26*, *29*
  administration/politics 55–82
  blue-green algae dynamics *119–20*, *122*
  drainage basin 25–54, *26–8*, *30–1*, 106–7, *107*, 122–5, *165*
  ecological modelling 116–22
  end-user information 133–9
  geographical features 26–9
  history 42–3
  ice-fishing *166*
  ICS 139–46
  institutional frameworks 158–60
  loading scenarios 121–2
  modelling scenarios 101–2, 156–7
  nitrogen input *115*, *117–18*
  phosphorus *116*, *121*
  physical environment 25–54
  practical recommendations 164–5
  qualitative scenarios 83–93
  quantitative scenarios 101–26
  river loads 34–40, *35–7*, **38**, *39*
  scenario results 125–6
  scientific information 153
  sediment reactive phosphorus *121*
  sensu stricto area 28
  simulated loads **110**, 111
  water quality 40–51, *40–1*, **40**, *43*, **45**, *46–7*, *49–50*
  web portal *145*, 146
Lake Pihkva 28–9
lake typology 155
land cover, Lake Peipsi basin *27*
land-use changes 32–4, 100, 153
Latvia 10–11, 56
legislation implementation 3–4
light limitation function (LLF) 120
Lithuania 56
LLF *see* light limitation function
loading scenarios 121–2, *124*
  see also nutrient loads

local government 8–9, 67–70, 142, 143
  see also municipal government
long-term hydrological fluxes 104

MacSharry plan 74–5
Mann–Kendall test **38**
MANTRA-East project 16, *16*, 20–2, 81, 150–2
map use 138, 139
mapping stakeholders 141–2
MEI *see* morpho-edaphic index
mining waters 32
Ministry of the Environment 67
Ministry of Natural Resources 58, 59
modelling 101–26
  calibration 108–10
  driving force changes **112–13**
  implementation/data input 105–6
  Lake Peipsi scenarios 101–2, 156–7
  nutrient loads 111–16, 164–5
  output 106–7
  performance 122–5
morpho-edaphic index (MEI) 44
multi-level governance 76, 78
municipal government 62–3, 65
  see also local government

Narva River 25, *26*, 28
National Environmental Action Plan (NEAP) (Estonia) 66
national-level stakeholders 130, 142
national power 80
National Research Council report 2–3
natural sciences 18–20, 22, 152–3
'naturalistic' view, social sciences 19
NEAP *see* National Environmental Action Plan
NES *see* Estonian National Strategy
network model 12–13, 129, 131
NGOs *see* non-governmental organisations
nitrogen emissions 107–8, **108**, **114**, *115*
nitrogen loads
  blue-green algae 164–5
  modelling 102–5, *109*, **110**, *117–21*
  riverine 37, 39–40
  scenario results 125
nitrogen losses *14*, 15
non-governmental organisations (NGOs) 130, 139
non-point pollution sources 32–4
  see also diffuse pollution sources
norms 11
Northeast Europe 8–9, 13
nutrient emission and transport model 102–3, *103*, 104–5
nutrient emissions 32–4, 103–5, 107–8, 111–16, 164–5
nutrient loads
  Lake Peipsi 34–40, *35–7*, **38**, *39*, 49–50
  modelling 111–16, 164–5
  scenario results 125
  water quality impact 121–2
nutrient losses 13–15

observed loads *109*, 123–4
oil-shale mining waters 32
open systems 19
oxygen depletion 28–9

phosphorus emissions 107–8, **108**, **115**, *116*
phosphorus loads
  blue-green algae 164–5
  Lake Peipsi 44, *48–9*, 50
  modelling 102–5, *109*, **110**, *119–21*
  riverine 39–40
  scenario results 125
phosphorus losses 32
physical environment, Lake Peipsi 25–54
phytoplankton 40, *41*, 45, *124*
piscivorous fish 44, **45**
point source pollution 29–32, *31*, 64, 104, 107–8, 165–6
Poland 75
polarities, integrated approach **18**
policy process
  agricultural sector 76–80
  Baltic States 12–13
  comparative studies 162
  EU changes 1–4
  institutional frameworks 158–60
  Russia 58
  stakeholder/public participation 163
policy scenarios 84
political control 76–8
Political Model (institutions) 158, **159**
political systems
  Baltic States 9–13
  communications 128

decentralisation 8–9
EU–Russian cooperation 7
information 150–2
Lake Peipsi region 55–82
practical recommendations 162
qualitative scenarios 87, 89–92, 96–7
water policy 2, 4
pollution
agriculture 7, 64–5, 70–2
Estonia 70
integrated approach need 15–16
Lake Peipsi basin 29–34
modelling 104
practical recommendations 164–5
Pskov Region 64–5
transboundary waters 4
population factors 27, 88, 90–3
positivist approach 18–19
power
information as 135
reform process 80
pressures, cause–effect chain 136
prevention strategies 49–50, 164
problem identification/solving 1–24, 80–1
Pskov Region (Russia) 30–1, 59–65, *63*
public
communicating with 127–48
information 140, 150–2
participation levels 140–1, *142*, 163
Putin, Vladimir 57, 58

qualitative scenarios 21–2, 83–99, 110–11, 156
see also quantitative scenarios
quality *see* water quality
quantitative scenarios 101–26
see also qualitative scenarios

Rational Model (institutions) 158, **159**
rational policy implementation model 13
RBDs *see* river basin districts
RBMP *see* River Basin Management Planning
recommendations 162–6
recreational fishing 89
reed belts 50
reference conditions 43–4
reforms (CAP) 71, 74–6, 79
Regional Fishing Inspectorate (Pskov) 61
regional government
European Union 8–9
Russia 57, 60–2, *63*, 65
regional scenarios 22
research groups 142
resource view of information 134
responses, cause–effect chain 136
river basin districts (RBDs) 153–5, *154*, 166
River Basin Management Planning (RBMP) 154–5
river basins 4, *5*, 13–16, 19–20, 25–7
river network nutrient transport 104–5
river nutrient loads 34–40, *35–7*, **38**, *39*, 102–10, 111
role game 134–9
Russia
administration/politics 56
agricultural sector 34
border regions water management 6–8
cooperation problems 7, 9, 11–13
environment management 57–65, *60*, 80–1
ethnic relations 10–11
Lake Peipsi basin *30–1*, 35–6
Russian–Estonian border 85–98, 132–9, 150–2
see also Lake Peipsi

SAPARD *see* Special Accession Programme for Agriculture and Rural Development
SAS model *see* story-and-simulation model
satisficing concept 158
'scenario fatigue' 86
scenarios 21–2, 110–11, 149–68
see also qualitative scenarios; quantitative scenarios
scientific differences 18–20, 161–2
scientific information 152–3
sediment reactive phosphorus (SRP) 120, *121*
semi-natural areas 27
set-aside scheme 74
SHALMOD ecological model 117–21, *123*
simulated loads *109*, **110**, 111, 123–4
Single Market principle 73
small catchments 123–4

social perceptions, water quality 50–1
social science approach 18–20, 22
socio-economic development 132, 137–8, 163–4
soil classification 27, *28*
soil system nutrient transport 104–5
solidarity principle 73, 79
source apportionment 114–16
source emission modelling 102–10
Soviet Union 57–8
  see also Russia
Special Accession Programme for Agriculture and Rural Development (SAPARD) 76
SRP *see* sediment reactive phosphorus
stakeholder mapping 141–2
stakeholders 127–48, 163
standard of living concept 93
State Research Institute of Lake and River Fishery (Pskov) 61–2
statistical trend tests **38**
story-and-simulation (SAS) model 22, 84–5
sub-national level communications 130–1
subsidiarity principle 75, 78–9, 80
summer water blooms 45–8, *47*

target scenarios *see* fast development scenarios
Tartu City (Estonia) *31*, 32
Tennessee Valley Authority (TVA) 160
'traditional' water management 17
transboundary cooperation *see* cooperation levels
transboundary relations 8–9, 11–13, 150–2
transboundary scenarios 84
transboundary water management
  future directions 149–68
  information for 150–3, 158–60, 162–4
  problematic nature 4–5
  tools 153–5
transport model, nutrients 102–5, *103*
trophic status, Lake Peipsi **40**
TVA *see* Tennessee Valley Authority

uneven distribution system (CAP) 78–9
USSR *see* Russia

Võhandu River *35*, *109*

waste management 63–4
wastewater treatment plants (WWTPs) 29, *30–1*, 62–3, 88, 92
water blooms 45–8, *47*, 48–50
water discharge *37*, **38**
Water Framework Directive (WFD) 2, 153–5
  EU border regions 7
  integrated approach need 16
  IWRM mismatches 161
  practical recommendations 166
  water quality 51
water policy *see* policy process
water quality
  economic development 34
  Lake Peipsi 40–51, *40–1*, **40**, *43*, **45**, *46–7*, *49–50*
  modelling 116–22, 156–7
  Russia 63
web portal *145*, 146
wetlands *94*
WFD *see* Water Framework Directive
winter oxygen depletion 28–9
WWTPs *see* wastewater treatment plants

Yeltsin, Boris 57, 58

zooplankton 41, 45

IWA Publishing's authorised EU representative for General Product Safety Regulations is Diane D'Arras, 15 rue Duret, 75116 Paris, France, e-mail: safety@iwap.co.uk.

Printed and bound by CPI Group (UK) Ltd, Croydon, CR0 4YY
18/06/2026
02142112-0001